Legacy of a Gentle Genius

The Life of A. M.O. Smith

Edited by Tuncer Cebeci

HORIZONS
PUBLISHING

Long Beach, California
Heidelberg

Legacy of a Gentle Genius

The Life of A.M.O. Smith

Edited by

Tuncer Cebeci

HORIZONS
PUBLISHING

Tuncer Cebeci
Aerospace Engineering Department
California State University
1250 Bellflower Blvd.
Long Beach, CA 90840-8304, USA
email: cebeci@csulb.edu

Library of Congress Cataloging-in-Publication Data
Legacy of a gentle genius: the life of A.M.O. Smith / edited by Tuncer Cebeci.
p. cm. Includes bibliographical references (p.).
ISBN 0-9668461-1-7 (hard cover: alk. paper)
1. Aerodynamics–Research–United States–History. 2. Smith, A. M. O.
3. Aerospace engineers–United States Biography.
I. Cebeci, Tuncer. TL570.L335 1999 629.132'3'092–dc21 99-24379 CIP

ISBN 0-9668461-1-7 Horizons Publishing Inc., Long Beach

Printed in Germany

Typeset by the editor
Cover design: Erich Kirchner, Heidelberg, Germany

Printed on acid-free paper 5 4 3 2 1 0

To A.M.O.

Our Boss, Our Mentor

who

- *hired us,*
- *educated us,*
- *guided us to excellence.*

Tuncer Cebeci

on behalf of

The members of A.M.O.'s
Aerodynamics Research Group

Preface

Often, one is unable to appreciate a rare opportunity until time has passed and history demonstrates its uniqueness. AMO Smith's Aerodynamics Research Group at Douglas Aircraft Company was such an opportunity. Those of us fortunate enough to be part of it enjoyed AMO's relaxed mannerisms "I am from Missouri - show me!", "Hum?" and his vision to choose the important topics to investigate. However, it was later that we realized how critical his guidance was and what an education it was to be a part of his Research group. We not only learned from AMO and the process but from each other. AMO was always partial to similes and I would liken AMO's group to a farm system in minor league baseball. As he hired each of us, trained us, and put us through the process, we finally became educated, specialized, and part of the majors!

In 1982, McDonnell Douglas established a parallel path to career development in the corporation. Similar to the management ladder where different management levels are recognized, the Fellow program recognized three levels of technical excellence; Fellow, Senior Fellow and Distinguished Fellow. Four people from AMO's original group of a dozen were recognized. I became a Senior Fellow in 1982, John Hess a Fellow in 1985, Bob Liebeck a Fellow in 1988 and Joe Giesing a Fellow in 1993. I was promoted to the Distinguished Fellow level in 1993 and Bob Liebeck to the Senior Fellow level in 1995. As of this date there were over 40 Fellows, 10 Senior Fellows and one Distinguished Fellow in the corporation of 70,000 employees in McDonnell Douglas prior to its merging with the Boeing Company. Considering these figures, one could easily state that AMO's farm system produced major players.

Other members of this group also had distinguished careers. Until their retirement from McDonnell Douglas, Andy Bauer was a major contributor to experimental work in aerodynamics and acoustics, Kalle Kaups in all phases of aerodynamics and Doug Smyth in aerodynamics design. Wayne Oliver is currently an established leader in design of high lift aerodynamics.

Outside the corporation, several of us have been recognized by professional societies for achievements and contributions in our areas of expertise. Bob Liebeck was elected to the very prestigious National Academy of Engineering in 1985, and AIAA fellow in 1990. Today, his concept of Blended-Wing-Body program represents a potential revolution in subsonic transport efficiency. John Hess made major contributions to inviscid flow calculations by developing the so-called Douglas Neumann code, a computer program that predicts the complete flow field of an aircraft. His work is often referenced in many technical papers and books. My work in boundary layers and turbulence modelling has also received recognitions and led to the Fluid Plasma Dynamics Award in 1984 and the Aerodynamics Award in 1993, both from the AIAA, and the Fellow award in 1986.

We prepared this book enumerating AMO's main areas of research in his career at Douglas from 1938 to 1975. Each area of his research and his technical papers is described by a member of the Group.

In conclusion, I would like to explain the choice of the title of this book, Legacy Of a Gentle Genius. In every aspect of the word, he was truly a gentle person. He was what you saw without airs or pretense. He probably would object to the description of himself as a genius. He was a pioneer in many aspects of aerodynamics and a visionary in recognizing the computational tools needed in aerodynamics for aircraft design. He was a mentor to many, a man of honor and dignity and a man dedicated to his profession.

February, 1999 Tuncer Cebeci

Contents

1

The Life of Apollo Milton Olin Smith

July 2, 1911 – February 2, 1997

BIOGRAPHICAL INFORMATION

Immediately after graduation from California Institute of Technology in 1938, Apollo Milton Olin (AMO) joined the El Segundo Division of Douglas Aircraft Company as Assistant Chief Aerodynamicist, the position he held until 1942. During this period he worked on aerodynamic and preliminary design problems of the DC-S, SBD, A-20, and DB-7 series aircraft among others. The last airplane upon which he worked during this period was the A-26 (later called B-26) attack bomber. On this aircraft he had prime responsibility for the detailed aerodynamic design.

In October 1942, at the specific written request of General H. H. Arnold to Douglas, he went to Aerojet Engineering Corporation on a leave of absence basis to help organize and develop the company as their first Chief Engineer.

He was asked to do this work because of his earlier work at Caltech on rockets. This earlier work led to the GALCIT rocket research project which later evolved into the Caltech Jet Propulsion Laboratory. For this early work he is commemorated, along with six others, in a bronze plaque at the entrance to JPL.

Under AMO's guidance the basic engineering organization at Aerojet was established and the engineering department grew from six people at the beginning to over 400 about 18 months later, when AMO returned to Douglas. This period saw the JATO type rocket through development and into quantity production. Like emphasis was given to liquid propellant rockets, but they never left the experimental state during this period. For his work at Aerojet he received the Robert H. Goddard Memorial Award of the American Rocket Society in 1954.

Upon his return to Douglas in March 1944, he resumed his work in aerodynamics and preliminary design. He was responsible for the detailed aerodynamics of the D-558-1 Skystreak research airplane which for a period held the world speed record. His role in this work is covered by the book *Supersonic Flight - Breaking the Sound Barrier and Beyond*, by R. P. Hallion, published by MacMillan & Company in association with the Smithsonian Institute. His design won the competition for a night fighter, which became the F3D-1 Skyknight.

At the end of World War II he was a member of the U.S. Naval Technical Mission in Europe spending about three months on the job. As an indirect consequence of this work he proposed and began studies of a tailless aircraft. These studies culminated in the F4D-1 Skyray interceptor which for a period held 6 official FAI World's records, one for absolute speed and the others for climb performance.

In 1948 he became Supervisor of Design Research, the position he held until 1954. During this period he conducted a considerable variety of studies, mostly with an aerodynamic bias. The work receiving the most effort was in the field of laminar flow control, but perhaps the most important accomplishment was development of a very general method of calculating low speed flow about arbitrary bodies. This method has become the well known Douglas Neumann Program or Panel Method.

In 1954 he rejoined the Aerodynamics Section as Supervisor of Aerodynamic Research, and his work was in essentially this same area until he retired. His work dealt with aerodynamic problems of a research nature. The main accomplishments were development of practical methods of analyzing laminar and turbulent boundary layer flow, new and improved static pressure probes, the hydrogen bubble technique of flow visualization, potential flow analysis, analysis of stability and transition of boundary layers, and introduction of the e^n method of predicting transition. He has pioneered in the application of electronic computers to many of these problems. Also he was a lecturer from 1954-1958 on Boundary-Layer Theory at UCLA.

From 1969 to 1975 he had the title of Chief Aerodynamics Engineer - Research. For his work in the development of potential flow methods he received the Douglas Engineering Achievement Award in 1958. He is the author or co-author of numerous Douglas Reports and over 60 external refereed papers.

In June 1975, he retired from McDonnell Douglas with no particular plans, but soon was appointed Adjunct Professor at UCLA for the period 1975-1980. In addition, he consulted to the Rand Corporation, 1975-1977, McDonnell Douglas (MDAC) 1975-1985, ARL at Pennsylvania State, 1977, Dynamics Technology, Inc. 1977-1981. In 1978-1979 and 1981-1982, he consulted to Bolt Beranek & Newmann; in 1979, to Pacific Sierra Res. Corp., the Foreign Technology Div., U.S. Air Force, in 1982 to Honeywell, Inc., and in 1985 to AMETEK. In addition, he has served on occasion as an expert witness in patent suits from 1977 to 1983. In 1985, he was a part-time Professor at California State University, Long Beach.

BIRTHPLACE

Columbia, Missouri

EDUCATION

El Dorado Junior High School, El Dorado, Kansas	1923-1925	
Hammond High School, Hammond, Indiana	1925-1927	
Woodrow Wilson High School, Long Beach, California	1927-1929	
Compton Jr. College, Compton, California	1931-1932	
California Institute of Technology, Pasadena, California	1929-1936	(BS in ME)
Califomia Institute of Technology, Pasadena, California	1936-1938	(MS in ME) (MS in AE)

PERSONAL

Residence:	2245 Ashbourne Drive, San Marino, California 91108
Telephone:	(818) 799-6173
Married to:	Elisabeth Caroline Krost (December 5,1943)
Children:	Tove Anne, Gerard Nicholas, and Kathleen Roberta

Amo's family photo in 1955. AMO, Tobe, Betty, Nicki and their dog, Toki

2

Some Autobiographical Recollections of A. M. O. Smith

COLUMBIA, MISSOURI; STILLWATER, OKLAHOMA, EAST ST. LOUIS

I have never kept a diary, so many of my recollections will not be completely accurate; but I hope you will find them interesting. As I am writing this shortly after my 83rd birthday, many of the early recollections will be rather fuzzy. Also, at times I will mention something that I was not involved in, but found interesting.

I was born July 2, 1911 in Columbia, Missouri. My parents were married on June 18, 1910. I do not remember anything about Columbia, but do recall a picture of the house, which was a California-style bungalow. In that picture was my father and me when I was about two years old. He was in overalls doing some work on our house, using a full-size wheelbarrow, and I had a miniature toy wheelbarrow.

My father graduated from the University of Missouri in June 1914 with an A.M. degree in chemistry. (See who's Who in the West, 1954). Then from 1914 to 1916, he taught at Oklahoma A & M in Stillwater. I have only two memories of this period of time. One time my father took me into the chemistry lab on a weekend. He was "monkeying" with some electrical circuits under a lab counter. All of a sudden he jumped back out from there, having gotten a shock. I thought the way he bounced back was *very* funny! The other memory was about a near accident. My mother was going some place with me, my sister Diana (about two years old at the time), and my sister Athena in a baby buggy.

Diana and I were walking ahead on a sidewalk. Coming to a cross street, we started to walk across. About that time an inexperienced woman driver turned around the corner and ran over Diana and me. The cars in those days had higher clearance than the cars nowadays, so she straddled Diana who got by without any injury. I was a little behind Diana and got caught by a wheel. It brushed against my face and "skinned" me up some, mainly one ear, but that was all. My mother saw it all, being behind on the sidewalk, and it really scared her. A few days later, the woman came to see how we were and brought a big box of candy. With that, my sister and I thought our accident was worth it! An inch or two further under the wheel and I would have been a goner.

From 1916 to 1918 we lived in East St. Louis, where I started school. We lived near the edge of town, next door to a large bare area, filled with weeds, which gave my father the idea to make a rabbit trap. We caught a number of rabbits and always had them for supper.

Usually I was like any other youngster, playing all afternoon. But one day I felt tired and went upstairs to take a nap. My mother could not find me, never thinking of looking upstairs, she had the whole block out looking for me. Then, to her amazement, I woke up and came downstairs!

At East St. Louis my father was night superintendent of what is now part of Monsarito Chemical Co. There was a bad race riot in East St. Louis at that time. Once we drove by a row of negro houses that had been burned down. I was told that the whites would set the houses on fire; the negroes would run out, and then the whites would throw them back in. During this stressful time, my father would drive to work armed with a 38-caliber pistol and a vicious looking dagger, but he never had any trouble so far as I know. But one day we were driving somewhere down town and suddenly we saw a white man chasing and shooting at a negro who ran around to the other side of a streetcar. What happened I do not know, but such things stick in one's memory. I think our car was a Maxwell.

One time we all went to the Anheuser-Busch brewery in St. Louis. On their tours they served free beer and pretzels. I didn't think much of the beer, but I liked the pretzels fine. Our house in East St. Louis had steam radiators. Once I must have been fooling around and fell with my head against one of the fins on the radiator. It was a strong blow, but not quite a concussion. My father, having worked in animal husbandry at the University of Missouri, knew what to do and dressed my wound, which was a rather bad gash. I still have a quite noticeable dent in the top of my skull.

SAND SPRINGS, OKLAHOMA

Next we moved to Sand Springs, Oklahoma, where my father had a job with a small chemical company, the Mid-Continent Chemical Co. (To know more about my father, see Marquis Who's Who in the West, 1954.) We first lived in a small house down-town where "Skeezicks" was born (my brother Orpheus Cecil, born 1918, died of diphtheria in 1921).

I don't remember much about this house, but nearby was a big livery stable. During 1918, one day it burned down, and later we went by to see the ruins. At least a dozen horses died. The fire had been so fierce that it burned through their abdomens and all their guts were protruding.

Soon we moved further out of town, to 419 Grant Street, where we stayed until I was through 4th grade. There was a boy about my age named Dennis who lived across the street. One day I saw him busy with a hammer and saw building something. So, I went across and asked him what he was making. His reply was that he didn't know; he would have to wait and see when it was finished! Behind us in the opposite direction lived the Towes. They had a big back yard fairly full of pigs and cows. In fact we called them "the Towes and their cows." The Towes were really dirty people. In summer, I was over there several times and they had their back door to the stable area open, and one or more pigs seemed welcome in the house.

This was the period of the World War I flu epidemic, and I remember seeing ambulances or hearses passing by fairly often.

I believe we still lived on Grant Street when I was in the 4th grade. Our teacher was Miss Williams with whom my mother became good friends. There seemed to be two smart kids in the class—me and Oleta Tillary. Often Miss Williams would have the two of us go to the blackboard, give us a substantial arithmetic problem and we would race to see who could get it done the fastest. She was probably doing this to help teach the other kids.

I was a pretty good kid in school, but not above some tomfoolery. Once Miss Williams was out of the classroom and I sailed an airplane across the room, and just as I did it, she came in. She called me to the front of the room, got out her nice oak paddle, made by some of the kids, and gave me a few whacks. It was winter and I was wearing two sets of underwear, so the whacks didn't hurt at all. I think I just stood there and grinned. The paddle was called the "board of education" by most of the kids.

Later we moved to 512 North Main Street. Now I went to a different school which was on top of a hill. At recess we often played shinny, a kind of hockey using a condensed-milk can as a puck and sticks properly cut from the branch of a tree or bush to resemble the shape of a hockey stick. Due to all the blows, the can would soon get badly beaten up with plentiful sharp edges. Once in a while it would hit a fellow and cut him. Another recess game was "tops". We would stand around a ring and spin our tops, which you hoped would drift out of the ring. If they didn't, they were fair game, and if you could throw your own top and knock the other out, it was yours. I was fairly good at this.

In those days tires were high pressure, so the inside diameter was fairly large. One sport was for a kid to ensconce himself inside the tire and then start rolling down the hill, roughly a hundred yards. It was quite a thrill to go upside down and round and round and miss the numerous trees, all at a good pace.

Up our street but at the edge of town lived a millionaire Indian, Charley Page, who was a benevolent sort. He had established an orphans home at the edge of town. Several of the orphans were in my classes at school. Charley Page also built a rather large amusement park near town. It had a lake, and on the Fourth of July he would put on a big fireworks show over the lake. In connection with the Fourth, I made a cannon out of a pair of 18-inch wagon wheels and a piece of 6-inch clay sewer tile. I had a 4-inch cannon cracker which I put inside and lit. It went off and the sewer tile disintegrated into a pile of shards. A the end of the school year Charley Page would throw the park open to all the school kids, and we would have quite a day. There was some kind of vending machine there on which you could play games and get a prize—some candy, I believe. I wanted the candy, got some other kids to put in their money and play the machine. Then I diverted them so they forgot about the candy. And I got my piece of candy!

The Sand Springs area was Osage country, and many of the Indians received royalties from oil. I noticed that at school you were not really "in" unless you had some Indian blood in you.

The Midcontinent Chemical Co. closed down and later my father took it over as the O.C. Smith Chemical Co. It was about a mile out of town, with a railroad spur running to it. Once in the summer between the 4th and 5th grades near noon I was taking some lunch and a bar of solder to my father at the plant. I was walking along the railroad and for no reason at all had slipped the 18-inch bar of solder up my sleeve. Pretty soon I noticed three colored kids stalking me, aiming to pick a fight. They kept getting closer and closer, and I didn't know what to do. I turned around and faced them, at the same time letting the bar of solder slip out of my sleeve. When they saw this "lethal" instrument, they just melted away. That evening I was downtown and one of these kids, now immaculately dressed as for a wedding or funeral, came up to me and apologized. I expect the negroes were afraid of trouble.

The people four house down from us had a cow, and I had a BB gun. So, we neighbor kids would often use the cow for target practice. An interesting target was the cow's udder. The cow was at least 200 feet away, so we didn't hit it very often, and we were too far away to do any damage. But once in a while we would get a bull's-eye; and when we did, she would twitch her udder; so it was easy to tell when we made a hit.

Sand Springs is on one side of the Arkansas River. While we were there they were building a bridge across it. One time my father and I went for a hike to the other side of the river. The bridge was only half finished. As I remember, 2" x 12" boards ran from one steel cross beam to the next, and were at least 50 feet above the river, with nothing but this single line of two-by-twelves between us and the river. We both got safely across and back, but I don't think I would let my kids try such a precarious path!

At the end of my fifth year in school, I went to Sunday School one Sunday and came home with a terrible pain in my right leg. By the next day, I was delirious. I had both phlebitis and typhoid fever. There was a blood clot at the top of my leg that blocked circulation. I had to lie very quiet so the clot would not break loose and stop my heart. This is certainly the sickest I have ever been. I was sick all summer, delirious the first month or two. My leg swelled up to about double its normal size, all full of fluid they called pus. In August, it was lanced and about a quart of pus ran out. I have forgotten the surgeon's name, but he had one quite withered hand. He had worked at length under x-rays, not knowing at that time the harm they could do. Of course, in the operation he used his good hand. I was kept in my father's den at home where the phone was located. Once at my sickest, probably in June, I seemed to be unconscious, and my mother called our Dr. Johnson. She said, "you better come over right away; he's dying." I was not really unconscious and said "huh?" This shook her up a bit. The den-bedroom was right beside the driveway. One day as I was recovering, several of my classmates came to the window and spoke to me. I had lost quite a bit of weight. With my first name Apollo, they coined the name Apilobones. In addition to the phlebitis, the typhoid just may me sicker and guided what I could eat. When I was recovering in August, the doctor would only let me eat ice cream. That was the one and only time I got fed up with ice cream.

EL DORADO, KANSAS

From 1922 to 1925 we lived in El Dorado, Kansas, where my father was Assistant Superintendent of the Skelly Oil Co. refinery. It was about a mile and a half out of town. There was a long row of company houses, with ours being next to the boundary fence of the refinery. So we lived more or less in the country. For the school kids, the company ran a small bus to take us into town and to school. A girl who was a year ahead of me in school Faye Palmer, would ride the bus, and in winter would always complain about being cold. But it was the "in" thing not to button your coat. So, I often needled her, "If you would button your coat, you wouldn't be cold."

Behind the houses was about 20 acres of abandoned pasture, giving us plenty of land. I put more than an acre of this into gardening, raising beans, corn, and mostly tomatoes. I had a least a half acre of tomatoes; nice large ponderosas, often weighing a pound or more. In fact, I grew so many that I set up a small stand by the road and sold them for $1.00 a bushel. I also had about 25 Rhode Island Red chickens. Often I would find large tomato worms on my plants. I would pick them off and throw them to the chickens. One hen would grab the worm by one end and run with it, followed by a considerable parade of hens going after the worm. After much chasing around, the original hen would lose the worm to another hen and the chase would begin all over again. I think the original hen could have eaten it right down, but the hens seemed to enjoy the chase.

My father told me to be business-like in raising the chickens. I did, keeping books on income from egg sales, expenses for feed, etc. When we moved to Hammond, Indiana, in closing out the chicken business after about three years, I found I had made $3.00!

Our life was pretty good out there in the country. My father had a 20-gauge shotgun, and he bought me a 410 gauge one. I often went with him into the woods near our house, hunting mostly for rabbits. A very easy way to get some rabbits was to drive slowly along some country lane at night. Rabbits would be on the road. When the lights of the car blinded them, they just sat where they were, being sitting ducks as it were. All you had to do was lean out of the touring car, fire, and go pick them up.

My father gave me a small wood lathe, which I put to a lot of use. El Dorado is on the Walnut River, and the main tree all along the river was black walnut. This was the most common native wood. As I have said, there was an abandoned pasture behind our house. All the abandoned fence posts were well seasoned black walnut, weathered to a beaufiful gray on the outside. If I wanted material from some wood turning on my lathe, all I had to do was walk a couple hundred yards and take one of these nice abandoned fence posts.

At one end of the row of company houses was a small pond that dammed up a small stream. The pond supplied water to the oil company. It was not potable, there being wells by the houses for drinking water. But the pond was a nice little lake, a few acres in area, with a concrete dam about ten feet high. In summer we would go swimming and the dam made a good diving platform.

Another memory was about Halloween. The company houses all had outhouse toilets. When Halloween was upon us, all the older kids living there would get together, go down the line and push over every outhouse. I don't remember any real damage. The next day when we rode down the main street in town, we saw that the kids in town had systematically taken all the garbage cans and emptied them in the middle of the street. Nowadays, all kids seem to do is knock on the front door and say "trick or treat". In 1925 there were plenty of tricks.

About 10 miles from our house was a tank farm with 55,000-barrel tanks storing crude oil. One time lightning struck one of the tanks, creating a tremendous fire. Crude oil tanks have a few feet of water in the bottom that settles out from the oil. If the fire burns down to the water, the water will boil and boil over; that is, foam up over the top of the tank and make a tremendous burning stream, at least ten times the diameter of the tank, even though there are dikes around the tank. We went over to see the fire at the time we did because my father knew they were going to shoot a cannon at it. We saw this happen, and a 3- or 4- inch burning stream ran out into the dike area.

My father gave me a clarinet which I gradually learned to play. He played trombone in the El Dorado town band, and got me into it too. I participated in several parades. One or more parades were for the Kafir Korn Karnival in El Dorado. Much kafir corn was grown in the vicihity of El Dorado; and in the fall El Dorado, being the county seat, had a fair by the above name. All the booths were extensively decorated with tassels of kafir corn.

El Dorado had two movie theaters, the El Dorado and the Eris. The El Dorado was the expensive theater, charging 25¢ admission. The Eris was a cheaper theater, charging only 10¢. Often on Friday evenings we would eat supper at the Owens family restaurant where you served yourself from large bowls of food set on the table; and then we would take in a movie at the Eris. My folks became good friends with the Owens who ran the restaurant.

A few years after the tragedy of Skeezicks' death in 1921, my parents "tried" again, and Hermes Orsino was born April 24, 1924 in El Dorado. Of course, with the difference in ages—I was 12, Diana was 10 and Athena was 8—when H.O. was born, he was the center of attention. We called him the "baby" for about a year; then "little kid", and finally Kidder until he was in high school.

HAMMOND, INDIANA

In 1925 my father was made chief engineer of the refinery division of the Graver Corporation in East Chicago. The company made and sold Jenkins Cracking stills, which my father helped develop at Skelly Oil. I entered Hammond High School some time late in the fall as a freshman. I do not remember many details of my schooling there. One thing I do remember was in connection with English. They gave three grades—scholarship, conduct and effort. In English I got an A for scholarship and C's in conduct and effort. I was rather proud of this.

I took two years of Latin there. Indiana had a statewide Latin contest for various grade levels. I was not much good at translating, but the grammar seemed logical and easy for me. In my second year I entered the contest, which was predominantly grammar. Hammond High School had been entering the contest since 1912, but had never even won at the county level. I won the county, went to the District at Valparaiso University and won that too. The finals were at Indiana University in Bloomington. My mother went with me. There, before the exam, I met some of the other contestants. They sure seemed brainy compared with me. Needless to say, I lost, but at least I had improved Hammond High School's reputation. In all this time, while I was winning, I still got only a B in Latin.

The school was rather strong in music and I joined the band. I soon became the solo clarinet. As the band entered contests, we went to different cities. One time I was riding to a contest with one of the leaders in his car. He came to the New York Central R.R. crossing which was at quite a slant to the road. He was too far to the side, a wheel caught the rail, and we ran up the track about 25 feet. Up ahead about a mile was the 20th Century Limited barreling down on us. As you guessed, since I am alive, our driver put the car in reverse and backed off in time.

Californians may not realize this, but back East in winter your spit freezed up, and in winter parades I would have about a 3-inch icicle at the end of my clarinet. Also, before any parade in winter, the director, Adam F. Lesinsky, would get together with all the trombones, trumpets, tubas, etc. and douse the valves with alcohol so the instruments wouldn't freeze up.

We had two interesting summers at Hammond. My father had to demonstrate guarantees, and train operators of the Jenkins Stills in Texas City, near Galveston. My father decided we should all go for the summer. He had a Chenango camp trailer, which folded out into two three-quarter beds, plus a stove, ice box, cupboards, etc. We lived in this trailer on the beach on Galveston Bay most of the two summers. We kids went swimming, fishing, crabbing, etc. These were good summers. The bay area near Texas City is loaded with mosquitoes. Our trailer basically had sleeping room for only five and I, being the oldest kid, was directed to sleep outside. I got badly eaten up by mosquito's, so bad that I became more or less immune to them. Now a mosquito bite will be gone in about 15 minutes.

At Hammond I got my first job. At the edge of town were many truck farms. I got a job weeding carrots for a Dutchman named Schoon. I did this, helped plant cabbages, worked on green peppers, etc. Nearby were about half a dozen Polish women who were working for another farmer. All this farmland was underlaid with clay tiles for drainage, which emptied into a river. The water leaving the tiles was nice and clear, and the Polish women were using it for drinking water. But somehow I found out it was full of small worms. I tried to tell the women about them, but they didn't understand a word of English. When I left this work, Schoon said if I ever wanted some more work to look him up.

When we moved to Hammond we stayed in a hotel for a week or two. I think it was the Lynwood. Then we moved to 121 Dyer Street, and after a year or so we moved to 1217 Harrison Street.

LONG BEACH, CALIFORNIA

In 1927 we came to Long Beach where my father had another Jenkins still to demonstrate. That fall I entered Woodrow Wilson High School as a junior. We first lived at 2429 East 4th Street. I again joined the orchestra, and played

Year 1929 — AMO in Wilson High School, Long Beach, California

2nd clarinet to Dick Russell. We both were heading for Cal Tech, and in my freshman year there, he was my roommate. Because I was fairly good at the clarinet, I thought some about becoming a professional musician. But I realized I did not have a lot of music in me and could see myself in some orchestra pit tooting away for some kind of accompaniment. This did not sound like a good life for me, and at the end of my junior year I dropped playing clarinet cold. I took a third year of Latin at Long Beach, and in the 4th year I took Spanish. With all the Latin I could practically translate Spanish extemporaneously.

Some time in 1928, Graver Corp. laid off my father. I think they were changing their plans about the cracking still business. So my father got a job with Western Oil & Refining Corp. in Wilmington, where he had been demonstrating the stills. We now lived at 265 Quincy in Belmont Heights, Long Beach.

In my senior year, I thought I would be a chemist like my father. I found high school chemistry to be "duck soup" and ate it up. I was program chairman of the chemistry club. My father told me about a stunt he had pulled off at a Lions Club meeting. It was to have a skilled speaker give a lecture about the chemistry of nitro-glycerine, including showing a sample to see what it looks like, emphasizing its sensitivity and danger. Then he fumbles the test tube and it is supposed to explode. We did this, using the high school debating champion as the speaker. I sat in the middle of the audience with a 38 pistol and a blank. When he dropped the test tube, I shot the pistol.

In preparing for Cal Tech I took some extra math. One fellow in the class was John Pierce who went on as one of the brainiest fellows ever to graduate from Cal Tech. Among other things, he has won the National Medal of Science. He is a member of the N.A.E. (National Academy of Engineering) as I am, so this math class had some pretty good people in it. In this class I got acquainted with John. Four of us, including Oliver LaRue, talked about building something, probably a sailboat. But a magazine article told about building a glider and had plans for it. This sounded more exotic than a sailboat, so some time toward the end of my senior year we started building it. We were flying it by early summer. In July, I was launched off the very top of Palos Verdes. I expected an updraft which did not come, stalled and landed with a crash before I reached the edge of the hill. The fuselage was a wreck but the wing design was not hurt. The fuselage design in the magazine was very poor; so John rebuilt the fuselage, both strengthening it and saving about 40 pounds of weight. The glider now weighted about 120 lbs, and I also weighted 120 lbs.

AMO and his gliders when he was in high school, 1929

AMO's glider when he was in high school, 1929

I could tell much more about the glider but will mainly mention a contest at Pacific Beach near San Diego, which Pierce and I both entered. About a dozen gliders were in the contest. The contest events were for Endurance, Distance, Altitude, and Spot Landing in two classes, sail planes and training gliders. Also, there was a sweepstakes cup. I won three cups—for distance, endurance and altitude. Pierce won one, for spot landing, which was interesting. The spot was marked by a blanket. The winner would be the one who stopped with the nose of the glider closest to the blanket. John did not quite get to the blanket; his glider was three or four feet away. One other glider stopped right on top of the blanket, but its nose was perhaps eight feet ahead of the blanket. So, John won.

We should have won the sweepstakes cup, which Hawley Bowlus won. The judges made a peculiar ruling. Two of Bowlus' flights were better than eight of ours; so the judges ruled those flights as winners, and he got the sweepstakes. My altitude win was peculiar. Launchings were with a shock cord pulled by about 20 Navy seamen. We knew our glider was strong, so we would take all the men could pull. The glider at launch was anchored, to a truck, and then released by a trigger device. The pull was enough to give an acceleration of about 4G so that the glider was off and flying in about four feet. Our control stick had a bicycle grip for a handle. In my winning flight, the acceleration was so great that the grip came off the stick, right in my hand. There I was climbing steeply with a free control stick. By the time I had recovered and gained control, I had won the altitude contest, which was determined by theodolite. The temporary lack of control is probably why I won this contest. People on the ground said I uttered a startled "gosh". That cup still sits in our library.

In my senior year I made all A's without ever taking a book home.

CAL TECH

I was accepted into Cal Tech and entered in the fall of 1929. I expected to be a chemist, but unlike high school chemistry, it didn't soak in at all, probably because I had not learned to really study. I did not do well, making mostly C's. So, at the end of year I got a job working at Western Oil & Refining where my father was superintendent, which helped me make money to go back to school. In the fall of 1931, since we had moved to 419 North Osage Avenue in Inglewood, I registered at Compton Junior College to try to get better grades, especially in math. I did well there; and when I reapplied at Cal Tech, they seemed glad to have me back.

About the time I first enrolled at Cal Tech, my father and I designed and built a much higher quality sailplane than our first glider, which was a primary trainer. The second glider performed so much better than on my first flight at Redondo. Nearly all I was trying to do was to bring it down. Most gliders had difficulty getting to the road below, but I easily cleared the road even though trying to bring the plane down, and finally ended up against a pile of sand about a mile further on, bashing in the nose. We decided we had a lot of learning to do and flew at various airports and flat fields. Later my father talked to James Irvine, and we got permission to fly on his ranch. One time I landed on a nice grassy meadow full of horses, which all trotted up and stood in a circle around the glider. I was worried that they might hurt something and waved the rudder at them. That caused them to trot off. We gradually got more experienced with the glider and decided to try the hills behind Corona Del Mar, where we flew a number of times successfully. Clyde White was a mechanic at Western Oil who had helped us with metal work such as brackets, etc. He also had learned, belonging to the Long Beach Glider Club, to fly a glider and had flown ours several times. On July 5, 1932, we were flying at Corona Del Mar again and offered him a ride. He took off, gradually gaining altitude for about a mile while the ground below fell away. Then, for some unknown reason, he put the glider in about a 30° dive; and when he was about

200 feet up, he put it in a vertical dive, killed himself, and thoroughly wrecked the glider. The CAA came out and gave us a clean bill of health. One of the few parts that still worked were the elevators. I often would think about this as I gained knowledge of aeronautics and never could see anything wrong in the design of the glider. White never seemed at ease flying, so maybe, as they say, he just froze to the controls. We piled all the junk on our trailer and hauled it to our new home in Inglewood.

When I returned to Cal Tech, I chose Mechanical Engineering as my major, hoping to take up Aeronautical Engineering later. There was no Aeronautical Engineering course for undergraduates—it was only a graduate subject at that time. I now did quite well in school. Cal Tech uses numbers for their grading system—4 as A, 3 as B, 2 as C, 1 as D. They also count all the time you are supposed to put in on a class, not just the class time. Thus, a full schedule was 52 hours or so. The last semester of my sophomore year I made 214 grade points. This was my highest, but generally I had a 3 average or better. Cal Tech divides up the school year into three parts. In the third part as a sophomore I signed up for an extra course, "The Theory of Equations." The class, being extra, had some very brainy people in it. I remember Max Millikan, the youngest son of Robert A. Millikan, who later became an economist, I think. Another was Kenneth Pitzer of the Pomona family. He later became President of Rice University.

In my junior year we were required to take 1) Business Law, 2) Economics, 3) Accounting. I knew I was doing poorly in Business Law, and after the final exam I went to see the teacher, Prof. Untereiner. He looked at his records and said I got a D by the grace of God. In Economics everything made sense to me and I got the highest A in the class. But in Accounting, I did not see what all the teaching was about. All I could see is if you spent something, you subtracted it; if you earned something, you added it—so I was back to getting another D.

I again stayed out after my junior year to earn some more money. I don't remember anything special about my senior year. By that time I was definitely interested in the aeronautics option. Mechanical engineers could take one aeronautics option course. It was taught by Dr. A. L. Klein, known as Maj., a nickname he got from his service in World War I ROTC. His class was mostly gossip about the aircraft industry which I ate up because of my glider experience. He was a consultant to Douglas Aircraft, so his gossip stemmed from real engineering problems. Later, in my second year of graduate school he taught a design course. In the course he gave us a mid-semester problem, and then a three-hour final. I got 100% in both of them; in fact, I was the only 100 percenter. But my final grade was a B. I went to see him about why I didn't get an A. His answer was that he didn't give enough written work to be sure anyone deserved an A. This left me frustrated.

When I entered graduate school, I got a job with the NYA (National Youth Administration) and was assigned to help Frank Malina prepare illustrations for a book being written by von Karman and Biot. Malina was a year ahead of me and a top notch artist who later became a very successful millionaire artist in Paris. Becoming a millionaire was due to not selling his stock in Aerojet which he helped found. When I worked with Malina, I found he was working on rockets. That sounded interesting, so I joined in, studying theory, working on tests, etc. My help was useful because of my mechanical engineering training and my experience with piping problems at Western Oil & Refining Corp.

Malina, Parsons, and Foreman had built a small rocket motor that they were to test in the fall of 1936. The test was in an isolated area above Devil's Gate Dam. I went along, helping move oxygen cylinders, sand bags, etc. I had a cheap explorer's helmet. Because they are supposed to be ventilated, one Sunday afternoon I made a real ventilator, using one of my mother's pepper boxes, tin snips, and a soldering iron. I put this on the hat. I was wearing it when we members of the rocket crew took a rest. Somebody took a picture of us, and I and my hat became sort of famous, or at least notorious. This original rocket firing in the Arroy Seco made people conscious of the area, and so was the reason JPL is where it is. At the entrance to JPL is a bronze plaque to this original group who fired the early rocket motor nearby. JPL has made a replica of this group photo which they display on special days, such as an open house.

For the von Karman and Biot book I made about 75% of the illustrations, with Malina giving me guidance. I learned a lot doing this work. Many figures are graphs of mathematical functions. We wanted the figures to be accurate, so I had to learn and understand the functions being plotted.

One time in my first year, 1936-37, Malina and I built a small rocket motor using alcohol and red fuming nitric acid. When we turned the valves on, the motor did not ignite and spewed very corrosive nitric acid. The fumes seeped all through the Guggenheim Aeronautics Building and turned all bright steel work such as calipers and steel rules very reddish with a special red rust. Our names were mud and we got kicked out of doing further testing in Guggenheim.

In the summer of 1937, Malina knew about a four-foot high cylinder of

Graduate students of Caltech's Guggenheim Aeronautical Laboratory directed by Von Karman, relax before testing their first rocket motor on Oct. 29, 1936, in the dry wash east of the present JPL. Schott, AMO in his self-designed helmet, Frank, Malina, Edward S. Forman and John W. Parsons.

compressed $N0_2$ downstairs under Gates, the chemistry building. We got permission to use some of it. The cylinder was old, being from WW I days, but we managed to crack the valve and some of the very noxious fumes started pouring out. Then, we could not get the valve completely shut off. These fumes are very noxious; so after several tries, all we could do was run away and leave the cylinder fuming. $N0_2$ is a heavy gas, and it just flowed over the lawn between Dabney and Gates, a nice lawn about 50 x 100 feet in size. The next day, that pretty green lawn had turned a rather bright yellow. Otherwise, we made progress in understanding rockets and what they would do. In the summer of 1937, Malina and I tried to find out some of the effects of air resistance on the motion of a rocket. This took step by step numerical calculation. I did most of this, working evenings at home on our sofa on Santa Ana Street. This work led to my first paper.

During 1937 I received an assistantship to work in the GALCIT (Guggenheim Aeronautics Laboratory-CIT, Inst of Technology) 10-foot wind tunnel. Bill Sears was boss of the testing crew. When I showed up for work, he looked at my name and saw A.M.O. SMITH. He repeated it and said, "what do they call you? AMO?" I said, yeah, I guess so; and that is how I got my nickname. Before that I was called Apollo, "Red" from the color of my hair, or sometimes AMOS.

The wind tunnel work was useful in many ways. The different companies would test models of their coming planes; so we got good previews of their plans. One model was of the Lockheed P-38 Lightning which became famous in WW II. Kelly Johnston did not like the performance of the original wing and took the model back to Lockheed to have a new wing built. Another model was of the Curtiss C46 transport. This model had more things wrong with it than any other model we had tested. Most were fixed during the wind tunnel tests, but it still was not a very successful transport. I will mention only a third model, but many were tested, averaging one every three weeks or so. This was the PBY, a Consolidated (General Dynamics) flying boat, that saw extensive service in the Pacific in WW II. Being a flying boat it had side floats to stabilize it. The designers at CONVAIR made the floats fold up and form a special wing tip or a sort of end plate. When we tested the model with and without these retracted floats, we found the drag was less with them on and retracted than off. So, they did indeed work as end-plates which definitely reduce drag. This finding was a pleasant surprise for CONVAIR. Another model that impressed me was the Northrop 7B, a hot twin-engine attack plane. It looked very good and influenced me to join the Northrop Division of the Douglas Company.

In my sixth year at Cal Tech I was asked to give a seminar about tailless airplanes. I read up on how they work and became familiar with the field. When it came time for the seminar, Northrop, Heinemann, his chief engineer, and John Wheatly, his chief aerodynamicist all showed up. I was scared to death, this being my first seminar, but I guess I didn't do too badly as later I got an offer of a job from Heinemann.

DOUGLAS EL SEGUNDO

I went to Douglas El Segundo Division in June 1938. I had five job offers, as the war scare was increasing. I was favorably inclined toward Lockheed, but they were dominated by a psychologist and couldn't make an offer until I passed some tests. I took them and passed with flying colors, but said "the hell with them" and went to Douglas. Incidentally, while I was standing around, one of their personnel men was contacting the applicants. In his hand was a letter from Clark Millikan about the Cal Tech prospects. The letter was about students going into industry. I spied on the letter and saw I was listed as number 1.

When I went to work at Douglas El Segundo, it was called the Northrop Division. Northrop had just quit. Later he was to found the new Northrop Company and asked to have his name back. Douglas gave it to him and changed the name of the plant from Northrop Division to El Segundo Division.

The first thing I will mention, I had nothing to do with, but nevertheless find it interesting. This Division had made many BT-l dive bombers. Shortly before I came to work, they began testing an experimental model of a considerably improved dive bomber, the XBT-2, with a more powerful engine, better wheel retraction, etc. This was the prototype of the famous WW II dive bomber, the SBD. The wing was exactly the same as on the BT- 1, however. The BT- 1 had good stalling characteristics, but the XBT-2 had unacceptable stall properties. My boss, John Wheatly, ran a number of flight tests, all to no avail. The only difference between the BT-1 and XBT-2 wings was in the paint scheme. The entire BT-1 wing was sprayed with aluminum paint, but later the Navy decided they wanted to have the top of their wings yellow and the bottom aluminum. Then one day Wheafly got an inspiration. Where the two colors joined at the leading edge was a slight ridge formed by the masking tape. Out on the flight line he got out his pocket knife and scraped away the ridge. He sent the airplane back up for flight test, and the vicious stall had been completely removed.

While I am on this subject of sensitivity to leading edge irregularities, I will mention an incident I was involved in—about 1950, I think. The F3D-1 was a jet night fighter that saw service in the Korean War. Once when undergoing flight tests, some of the sheet metal that helped to fair in the landing gear when it was retracted came loose when the wheels were extended during a dive. The fairing came off and flew away over the top of the wing, hitting it slightly. After repairs were made, a test pilot named Brownie put it through some flight tests. He said the plane was not quite right—it had a nibble at the tail, as he put it. The mechanics and engineers carefully checked over everything such as aileron rigging, but the nibble was still there. One Friday several came to see me to see if I had any suggestions. I went to the airplane, looked it over carefully and noticed one thing. The leading edges of these airplanes were protected by an approximately 1/16" thick coating of a rubbery material to protect against erosion by sand or rain, and where the fairing had hit the leading edge was a 6-inch long gash, about 1/32" deep. I told the crew to smooth it over and flight test it Saturday. If the trouble was not fixed, I would go up with the pilot on Monday and see what I could see. When I came to work Monday, they said the trouble was all fixed; so I missed out on an early jet flight.

When I first came to work at El Segundo, a secret project was a twin engine tailless bomber started by Northrop. I guess because of my seminar I was asked to handle the wind tunnel tests at Cal Tech. The tests first looked pretty good, but when results were corrected for trim effects, the performance was no better than that of a conventional bomber. On a tailless airplane, the elevons have to be turned up to get trim. This ruins the efficiency of a wing. So, when we got through analyzing all the tests results and went over them with Heinemann, he dropped the project. I am sure the airplane was mostly Northrop's idea.

A few comments about the DC-5 are in order. At El Segundo~permission was obtained to build DC-5, supposedly an improvement to the DC-3. It was certainly faster, about 220 mph, 180-190 mph for the DC-3. At this time, it was felt that the airplane was conventional enough so that no wind tunnel model was needed. (The DC-10 spent about $17 million on model building and $17 million for testing costs.) A number of DC-5s were manufactured and delivered, notably to the Marines; but with the war getting imminent, Donald Douglas said that unless sales of 500 could be guaranteed, the project should be dropped. It was.

Gene Root was the expert on tail design at Douglas. He was located in the Santa Monica plant. Heinemann contacted him about designing the tail surfaces for the DC-5. One day about quitting time, he came over with his proposed drawings of the tail. It looked quite large, and Heinemann and Root argued very stubbornly, both being very stubborn fellows. I stood behind them for about 45 minutes, while Root really stuck to his calculations. Then I happened to notice that the El Segundo drawings and the Santa Monica drawings were not to the same scale. One was 1/16 size and the other 1/12 size, I believe. When I pointed this out, Root's tail size looked OK. I prepared the official report of estimated performance of the DC-5. I estimated the high speed to be 220 mph. In flight test it came out to be exactly 220 mph. The other items like take-off distance, rate of climb, etc. also came out close but not quite so perfect. Probably this accurate prediction got to top management's attention.

Donald Douglas had a practice of inviting promising young new hires to his home for dinner. One time he was entertaining a Navy Captain who was the El Segundo Plant representative. I was invited. Everybody had to wear tuxes,so I bought my first tux. There were about a dozen people at the dinner, and I sat next to the first Mrs. Douglas. She talked about her three children. Two of them she felt pretty proud of, and then there was Junior. He had flunked out of Stanford and now was taking an engineering course at Curtiss-Wright Tech. She did not know what would become of him as he really had not found himself. So, because he had not found himself, he became president of Douglas Aircraft. Ironic, huh? Incidentally, I knew Junior almost right away. When he was at Curtiss-Wright Tech, he came to me for help several times on aerodynamic problems.

When I went to work at El Segundo, their main business was selling export versions of the Army A-17A attack aircraft, called the 8A-"something", such as 8A-3P for Peru and the 8A-3N for Holland. Some were being exported to Iraq. All the girls in the engineering department were totally fascinated by the very virile looking, swarthy Iraqian pilots. I was flight engineer on the 8A-5 destined for Norway. The flight engineer rides in the back seat and records all sorts of data from the instruments, like air speed. This airplane had a 2-speed supercharger, so it could fly higher than most. Once on a full power climb we got to 32,000 feet, and I held the record for going the highest of any civilian in Douglas Aircraft. On this flight, because of the low outside pressure at altitude, the oil started foaming and covered the windshield. Bob Lacey, the pilot, radioed in what was happening. Fire trucks met us at the end of the runway. After we reached maximum altitude, my work was done, so I just sat back and enjoyed the scenery as we descended. The oil only impinged the front windshield so I had no idea we were in trouble until we landed and saw the fire trucks. I felt sort of gypped. On subsequent flight tests to fix the trouble, if it occurred again, Lacey would radio back, "struck oil again". So, this high altitude place got to be known as Lacey's oil field.

Another adventure with Lacey was interesting. At that time, to calibrate the air speed installation, one flew at a low altitude (like 100 feet) between locations whose distance apart was known. We were to fly between the Huntington Beach Pier and the Newport Beach Pier. On the way down, we stopped at the Naval Reserve Building at the Long Beach airport to pick up some life preservers. Then we flew at various speeds between the piers with me holding a stopwatch. Lacey remarked that he must remember to switch fuel tanks because the flight was fairly long and at high power. After 5 or 6 runs the engine was cut back to idle and we zoomed up to more altitude. From the back seat all I could see of Lacey was that he was working his arms rather vigorously. The intercom was turned off. We slowly descended from about 1000 feet to about 100. I wondered if he was working the wobble pump. It began to look as if we might ditch the plane, so I gathered up my notes and got prepared. Then when we were at about 100 feet, the engine went back to full power and away we went. When we got to Long Beach, I asked him what happened. He said he had cut back the engine and was flying slow so he could reach out and wipe oil off the windshield.

In the fall of 1939, my boss, John Wheatley, quit and went to Northrop. We were scheduled to run high speed wind tunnel tests of the A-20 at NACA Langley in their 8000 hp wind tunnel. So, it fell to me to go there. Our model had a cast aluminum wing. NACA people at that time were not very receptive about outsiders using their facilities. So, when I got there, I entered an office full of people who were in a rather defiant mood. One of them said, "Your model is no damn good; what are you going to do about it?" This was quite a reception for a young squirt like me. We talked some and what had happened was that at El Segundo after the model was finished and met accuracy requirements, a tool clerk was given a buffing wheel and told to shine up the model. Cast aluminum does not take much of a shine; so this fellow laid into the wing and wore it down by 1/32" in places.

The NACA people told me about an outfit in Richmond that did metal spraying to build up the wing, as well as about a shop in Norfolk that could file it down afterwards to get correct contours. Mr. Roche was the Army representative, and we inquired about shipping the model to Richmond. The NACA people were very uncooperative and would not even lend us a truck to take the model a few miles to the railroad station in Hampton. So, Roche got mad, put the wing in the back of his Lincoln Zephyr, and we drove to Richmond, stayed while the model was metal sprayed, and then took it to Norfolk for accurate recontouring. This took several weeks, and I would commute back and forth on the ferry from Old Point Comfort to Norfolk to see how things were coming. When we finally got into the wind tunnel, everything worked out all right. I think about 8000 of the various versions of this airplane were built during World War II.

John Stack, a famous aerodynamicist, was the boss of the wind tunnel. He was a chain smoker and always tried to chisel smokes off his staff. Since his staff was mostly nonsmokers, about all he did was cuss them out for being a bunch of pansies. As that area has a strong southern accent, in order to be understood over the phone, I had to cultivate a southern accent.

In 1939 the B-23 was first flown. It was a cleaned up version of the B-18, which was a modification of the DC-3. The B-23 also had much more powerful engines. The Santa Monica aerodynamicists played conservative in their performance estimation, assuming the airplane was just as dirty as the B-18 and allowed only for the increased power in their high speed estimation. But actually it had been cleaned up considerably. However, the aerodynamics department policy was not to assume cleanup until it had been demonstrated. When it was flight demonstrated, its high speed was 282 mph as against their guarantee of about 260 mph. This is a tremendous miss, and I understand that the low guaranteed speed was the reason Douglas lost out to North American who was building the B-25 (Mitchell) light bomber. This got the Santa Monica preliminary design department so down on their own aerodynamicists that for the next year or so on any new preliminary design, they would privately ask me to run my own estimate of performance to check on their own aerodynamics department. Carlos Wood, head of preliminary design at Santa Monica, was the one who asked me.

The A-26 was meant to be an improvement of the A-20, cleaned up, with higher power engines, better armament, etc. I had a lot to do with its design. I remember choosing the wing plan form, the airfoil section, the nacelle design, the oil cooler inlet shape, etc. It was a very successful airplane that first flew in July 1942. About 2000 of this airplane were built. One thing sticks in my mind. The airplane had power driven machine-gun turrets, designed by G.E. Early in the airplane design, they came to us with a proposed exterior design of the turret. It was like a pill box you sometimes see in ground gun installations. I looked at it and told them and my boss, Heinemann, that this design would knock 40 mph off the high speed of the airplane. The G.E. fellows went back to their drawing board and came up with a much improved design.

I was at Aerojet from the fall of 1942 to March 1944, which will be discussed separately. By the time I returned from Aerojet, Heinemann had seen quite a few bad accidents. The pilot usually got into trouble when the airplane was flying the fastest and going straight down toward the earth. He expressed the wish to have an experimental airplane that could fly level and equal the speed that others reached in a dive. This somewhat naive notion was the start of the D-558. I worked a lot on the design of this airplane. Conceptually it was a case of wrapping the smallest airplane around the largest jet engine that was available. I chose the wing airfoil section and plan form. Also, I did a lot of work on the inlet design. (When I went to Aerojet, I had not heard of jet engines. When I came back to Douglas, they were a new gimmick and several models were available). Much of my time was spent analyzing and understanding compressible flow which accompanied the jet airplane era. I wrote a 150-page report about all its nuances. It was a regular report but was not entered into the Douglas report system, mostly because we just thought we were catching up with what other people like Pratt & Whitney, G.E. and NACA already knew. But as time went on, every so often I would see a newly published NACA paper that was on what I treated in one of my chapters. I kept this report for many years, but when I retired, I could not find it. The D-558 flew in 1947, and for a while held the world speed record of *650* mph.

AERO-JET

In July 1942, General Arnold wrote Donald Douglas asking for my services at Aerojet, a newly formed company. I saw the letter but never obtained a copy of it; and later when I tried to find it, I couldn't. The request was due to my rocket work at GALCIT. Douglas gave me a leave of absence of 18 months. So I went to Aerojet in September or October of 1942. Aerojet's job was to make the rocket hardware that J.P.L. was developing. When I came to work there, there were about a half dozen people in the engineering department. It was war days and these draftsmen seemed dead on their feet, putting in about 100-hour work weeks. My first official memo was to say that they had to get my permission if they were to work more than 60 hours per week.

When I first came there, the main work was to make rocket-assisted take-off units for airplanes using red fuming nitric acid and analine. Red fuming nitric acid is very corrosive, and with analine as a fuel, it is hypergolic; that is, as soon as the two come in contact, they burst into flames. So it seemed to me from my Douglas experience that it was far simpler just to lengthen a runway with a bulldozer or use steel tread plates which were in mass production during the war. It seemed to me that Aerojet was founded on faulty premises. But early in the game, one of the founders, Jack Parsons, said they were having some luck with a new kind of solid rocket that was a mixture of tar and potassium perchlorate, $KCLO_4$. These two materials are relatively innocuous, so I started pushing this development. By the time I left Aerojet in March 1944, the place was in mass production on such JATO (Jet Assisted Take Off) units that were fastened to the sides of airplanes for assisted takeoff. The engineering department had grown from the original six to more than 400; so you can see I was rather busy.

I could write a lot more about my Aerojet experience but won't. During the war Aerojet's product was secret. A common rumor around Pasadena was that they were making some kind of new carburetor, guided by the name. Later, I remember Col. Joiner, the Army representative at Aerojet at about the time the V-2 started into use, saying that the main reason for keeping our work secret from the Germans was to hide our ignorance from them. Aerojet was very unstable and crazy in their actions, and I got fed up. So, I went back to Douglas in March 1944, having overrun my leave, but they took me back. I guess I did all right. Haley, the President of Aerojet when he was nominated to be president of the American Rocket Society, told me that a condition he had made for accepting the nomination was that I be given the Robert H. Goddard Award, which I received in 1954.

MORE DOUGLAS EL SEGUNDO

When I went back, one of the main things I worked on was the D-558, as already mentioned. In the U.S. there was negligible high Mach No. data from model tests for estimating performance of the full scale article. In fact, the main test data was some German info I dug up on some rectangular brass airfoil models that were 2 cm wide and 10 or 13 cm long. I did performance estimates for the D-558 mostly based on this meager data. Two or three years later, one of the engineers came and showed me a plot of D-558 flight test drag data. It very closely agreed with my original estimate. In fact, that is why he showed it to me. Later when I was in Germany, I ran across these small airfoil models and took them home to Douglas. I accumulated several other wind tunnel models and formed a miniature museum.

Gene Root, my boss, and I were selected to go to Europe as part of a Navy mission to learn what the Germans had done technically during the War. The mission was called NAVTECMISEU and was headed by a Commodore, the equivalent of a Brigadier General. Root and I left the Burbank Airport on May 20, 1945 for Washington, D.C. We were still on the Douglas payroll, but the Navy simply paid Douglas our salaries.

We spent several days in Washington getting indoctrinated. One thing the Navy people warned us about was booby traps. The German would put some kind of explosive in a hollowed out book. Then, engineers and scientists, being curious about books, would open it and, bang, it would go off. We also had a walk through a woods that had been filled with would-be land mines, buried in the ground. For this exercise, the so-called land mines were large fire crackers. I never set any off, but my partner Root did. He said it stung his foot a little. In getting outfitted with our uniforms, I found myself in the same change room as Gene Kelly. So, I can say I took off my pants with Gene Kelly. We both were Naval Technicians, but he was going on quite a different mission.

We took off at midnight from Patuxent for the Harmon Air Force Base in Newfoundland in an R5D airplane, the Navy version of the DC-4. We stopped there about 6 a.m. for breakfast. Coming from coupons and rationing of food, it was one of the best breakfasts I ever had. Long tables were loaded with one-pound cakes of butter, big platters of bacon, fried eggs, fried potatoes and toast. It was a real contrast to our frugal war-time fare at home. After eating, and refueling the plane, we took off for the Azores, arriving there for supper. Then we took off again, flew all night, and arrived in Paris the next morning.

Root and I did not know what we could do or be assigned to do, but very quickly we found a useful job. At Göttingen, one of the major German aeronautical research facilities, when they found they were losing the war, they threw all their classified reports down a well. Our Army discovered them, recovered them, and brought them to Paris to dry, using the NAVTECMISEU Building which was next to the Etoile, where they spread everything out along tables in the 7th floor attic. Our job was to index this material, copy it and get it to the U.S. fast because the war with Japan was still going on. We had micro-filmers copy the material 24 hours a day, 7 days a week once we got going. Easily the most startling thing technically that we found was the swept wing. Except for some preliminary theory by R.T. Jones, no one in the U.S. had thought of swept wings, and of course no tests had been done. The Germans had tested hundreds of models.

The Navy always does well by itself. Our hotel was the Royal Monceau, a really deluxe hotel nearby. At dinner time, they always had a small orchestra. Lindbergh was also a member of this same mission, but was about to go home. The second day we were there we spent the evening in his room learning what he had found out. At the end we all gave him dollar bills to sign as "short snorters", then a custom for anyone who had flown across the ocean. He was more or less the original short snorter. We keep the bill in our bank vault.

The mission had several sections—electronics, aerodynamics, power plant, structures, etc. Our boss was Cmdr. Dayton Seiler. He was a very unusual Navy man. Clark Millikan of Cal Tech flew across with us and accompanied us at times. Seiler had been a Naval Postgraduate student at Cal Tech, and had stayed at Clark's father's home, the Robert A. Millikan home. Seiler seemed to be a regular Navy man, but was strongly religious. Clark said that Seiler would get on his knees and pray for Robert's and Mrs. Millikan's souls, sometimes for a half hour. The senior Millikans were not sure they liked it.

Paris was extremely short of common supplies, but they had plenty of perfume. I would deal with the hotel manager, and since I did not smoke, I had plenty of cigarettes to offer. I would trade a package or two of cigarettes, or some times a bar of soap, for an expensive bottle of perfume such as Chanel No.5. There was another mission in Paris, the ALSOSS mission. We often saw some of its people but never could find out what they were there for. When the A-bomb went off in Japan, I knew.

My wife said all we seemed to do for amusement was go to girlie shows, like the Folies Bergere. But we worked seven days a week, and all the museums were still closed; so there was not much else to do. The girlie shows were running full tilt. Once I went to the Paris Opera and saw Thais, so it wasn't that I didn't have interest in other things. From our 7th floor attic we had a fine view of the Arc de Triomphe, the Etoile, etc. American pilots were fairly restrained, but the British pilots were crazy. They really liked to buzz the Champs Elysee. From our 7th floor we could look down on them, and they even had to climb to clear the Arc de Triomphe. We saw a number of doings at the Arc de Triomphe from our vantage point, such as Eisenhower getting a medal from de Gaulle. We could see all the roof tops, being so high. On some of these occasions we could see soldiers manning machine guns on the roofs.

Later, to round out our work, we took a short trip into Germany to see what it was like. Germany had some fine autobahns at that time. In the U.S. at that time, the only equivalent one was the Pennsylvania Turnpike. But the Americans had done a bangup job of destruction. There was not a bridge left all along the several hundred miles we drove. We had to use temporary bridges put in by the military to get across.

I spent several days at Göttingen learning about the place. Most of the technical mission guys were in the field to learn about facilities and the like. Root and I were homebodies handling all the documents; but I did do two things out in the field that helped. One Sunday morning Col. Dane, the officer in charge of the occupying group at Göttingen came to me and said a German had shown up claiming he was a scientist, and would I evaluate him for possible transfer to the United States. The man was in a business suit complete with tie, but the dirtiest and most unkept you ever saw. He said he was Dr. Bernhardt Goethert. From all my work with the documents in Paris, I knew the name well. He had been in charge of their large wind tunnel in Berlin and had published many reports of importance to the U.S. war effort. He had previously sent his family to Göttingen. When the Russians took over, he was told just to stay at his home. One day he got word to report to Russian Army headquarters. Instead, he started walking to Göttingen. He walked 26 days, swimming rivers and overcoming all sorts of difficulties. No wonder his clothes were a mess. I wrote a memo to Dane saying he was one of the five most important Germans in the aerodynamic field to the American war effort. He was brought over to the U.S. and led a long and distinguished career for the Air Force and the Arnold Engineering Development Center (AEDC) in Tennessee. I saw him every so often and he always remembered me. So, this was one of my field contributions.

The University of Göttingen has one of the major libraries in all Germany. I went to see it, but since it was almost entirely empty, I was very disappointed. I had my own jeep driver, and that evening told him about my disappointment at finding the library empty. (My jeep driver was a happy-go-lucky ladies man. By the afternoon of our first day at Göttingen he had gotten acquainted with a Junior honor student at Göttingen. Her family even had a swimming pool; so when he was off duty, he lazed around there.) The next day he told his new girlfriend about my dismay. She readily told him what happened. When the Germans were worried about Göttingen being bombed (it wasn't), the entire student body turned out and carted the books a few miles away and stacked them in a tunnel alongside the railroad tracks that led into a salt mine. I passed this information to Dane, but didn't get to the mine because I was scheduled to leave. Later I did see him and he said they found all the books alright—in first class condition because they had been stored in a dry salt mine. This was the second of my field contributions.

Later we stopped at the V-2 factory at Nordhausen. It was under a hill that had earlier been a rock quarry. They had tunneled it out to make big rooms. It employed 26,000 people, and the railroad ran right into the hill. I could see many D.P.s around the entrance. Many seemed to be living in rejected tail sections of the V-2s. On my second day there I started at one end to explore the facility. I started at 9:00 a.m. and walked steadily until I came out the other end at 1:00 p.m. In the first part were all kinds of gauges, micrometers, and other precision instruments, which had been kept on many shelves. But when the war ended, the slave labor went crazy and pulled all this shiny precision equipment off the shelves and on to the floor. So I found myself walking on a foot deep layer of fine instruments. I took some of the instruments and still have them.

Soldiers seem to very casual about their lives. When I finished my walk, being after lunch, I went into the kitchen and got some lunch from the mess sergeant. We sat and talked. He said he was not feeling too well. Many of the D.P.s had gotten into what he called the V-2 juice, which was partly wood alcohol, and died. He wanted to know what it was like, so he drank some too! Later, in Braunschweig I encountered an Army captain in the decontamination squad whose job was to disarm bombs, etc. Most American bombs had what was called an anti-tamper fuse; so if you tampered with it, the bomb would go off. But this captain said they don't always work. He had just recently won a quarter disarming a 500-lb. bomb!!

Nordhausen had a crematory up on the side of a hill. With my jeep driver we went up to see it. On the downside was a large truckload of ashes. My jeep driver was just a happy-go-lucky seaman, but after seeing this, he said he hadn't really known why he was in the war, but now he knew.

My next stop was at the L.F.A. laboratories (the Herman Goring Laboratories) near Braunschweig. It was a tremendous laboratory, all hidden in the woods. Until the end of the war the Air Force didn't even know it existed. For instance, I went on the top of the three-story building containing a wind tunnel. The building was all made of poured concrete including the flat roof. The roof had been landscaped with grass, weeds, and good-sized trees to hide it. The visitors quarters were nice and located in a rather dense woods. At night I heard my first cuckoo calls—they imitate the clocks rather well. In the field of aerodynamics, the Germans were far ahead of the U.S. The wind tunnel in the building I just mentioned was powered by a 30,000 hp motor. In the U.S. the most powerful wind tunnel was 8000 hp. It was the tunnel I tested the A-20 model in. Furthermore, Germany had 20 supersonic wind tunnels of varying quality and sophistication. At this time, the U.S. had only one.

Later, we drove through Magdeburg, which was really only a pile of rubble, to a salt mine. I had been told they were making airplanes down in the mine. The building for handling the salt was certainly just an ordinary looking warehouse. The factory was about a thousand feet down, according to its superintendent who showed us around. There were big rooms, about 100 x 300 feet for the various stages of manufacture. The quarters were ideal in many ways, cool and dry because everything was salt. I remember wetting my finger and rubbing it on a wall. It was straight salt. The superintendent showed me the stockroom where I took a compass and an altimeter. The compass is now in my Ford. In the automatic turret lathe section everything seemed to be nicely in order, so I threw a switch and the lathe started turning out more parts. They did not make complete airplanes here, just fuselages. They would hoist them up the elevator shaft at first. When I left, I set the altimeter for the bottom and when we got to the top of the shaft, it was reading 300 meters; so I knew the superintendent was at least telling some truth. All the Germans I dealt with struck me as good sports. Once licked, they seemed ready to shake hands and call the game over.

Root had mainly gone to different areas than I had, but later we learned that the Swiss had been doing some topnotch research in shock-boundary layer interaction; so we made arrangements to visit them in Zurich at the Eidgenossichen Technischen Hochschule (ETH). The Swiss were neutral during the War, so combatants could not go in there in military uniforms. But we had shipped our civilian clothes back home. Therefore, we went to the U.S. Embassy where a tailor took off all our braids and stripes. People then said we looked like a pair of preachers. We had a very interesting visit both at the E.T.H. and also at some of the major factories. We talked to Prof. Ackeret, the famous head of the Aerodynamics Department. After we had talked for a half hour or so, he asked if we would like to see the wind tunnel and test set up. We said "yes" and he took us down. He, Feldman and Rott were doing the experiment. Rott is now a retired Professor at Stanford. In the process of looking at and talking about the experiments, Root had some private conversation with Feldman. After seeing the experiment Ackeret said "would you like to go this way?" He took us to a door, opened it, and we found ourselves out on the street in bright sunlight. It was quite a dismissal. Ackeret was mainly sympathetic to the Germans as he had many good friends there. But we got in touch with Feldman whose wife I believe was a member of the American Dietzgen family, a noted maker of drawing instruments. Feldman hoped to get to the U.S., so he treated us well. In fact, that night he took us to a fine restaurant. He had his research nearly written up as a Doctor's thesis. The next day we met him at his home where he let us spread all the pages on his dining room table and take pictures to send to the U.S. The table had a lace doily on it and our pictures gained a kind of fame in the U.S. with the doily background.

Switzerland had an abundance of products such as watches that they couldn't ship out. There were tremendous stocks. I came back to the U.S. with five or six watches.

BACK AT EL SEGUNDO

I do not remember exactly when I got back to El Segundo, but it was a week or so before Japan surrendered. Soon, R.G. Smith and I pretty much worked out the design of the F3D Skynight. Heineman was too busy with the AD-1 Skyraider to pay much attention to this competition. We submitted a design with two jet nacelles under the wings like a Boeing 737 transport. I had become aware of the ME262 which had the same arrangement; so we proposed it. With that arrangement it is fairly easy to change to different engines since they are exposed, under the wing; so this arrangement is quite adaptable. Well, we won the Navy competition. Then Heineman started paying a lot more attention. He did not like our nacelle arrangement and completely changed it to a pair of engines buried in the fuselage. I was not very happy with this casual overriding and started getting out of preliminary design. This airplane first flew in March 1948. The first AD-1 flew in March 1945. Earlier in this writeup I mentioned about missing a ride in the F3D.

Based a good deal on what I had learned in Europe, it looked to me like a very good design could be made out of a low aspect ratio, swept back, tailless airplane. It had a number of advantages that higher aspect ratio tailless airplanes like Northrop's lacked. Accordingly, one day I submitted a memo to Heineman with a sketch of a proposed wing plan form. He got interested, and I handled preliminary wind tunnel tests of the wing I suggested. The testing was done at the Cooperative Wind Tunnel (CWT) in Pasadena. The tests turned out well. I will mention one odd-ball observation. The CWT was carefully designed to handle tests of any model that might come along. There was much discussion of loads that the balances could handle, both maximum and minimum. The designers thought they had made a system that could handle anything. I came into the tunnel with two models, one 8- foot span for getting general data, especially stability data. The second model was a one-foot half span model that was to set on the floor balance. The big model had an area of 32 sq. ft. The small model had an area of 1 sq. ft. In the original planning of the wind tunnel, no one had ever thought of testing such a low aspect ratio wing. With the usual wing aspect ratio of 6 or so, the wing area of a model would be about 10 sq. ft. instead of 32 sq. ft. So, Clark Millikan, the Director, told me that with the big model I would exceed the design strength limits of the balance system. The small model was made to get high Mach Number data, and it was so small that its forces were below the sensitivity of the balance system. So with this one new airplane model I went beyond the design limits in both directions.

The model looked very good, but after handling the early design questions, I did not have much to do with the final design work. Four hundred twenty-one F4D Skyray fighters were made. The airplane first flew in January 1951 It held quite a few world records for speed and time to climb..

After this I got more into aerodynamic research. The area rule involves calculating the flow about a bumpy body of revolution. Existing methods for calculating it were very poor. K. E. Van Every, my boss, talked to me about looking over the available methods and seeing what was best. I did and discovered an entirely new method that was far more powerful than any of the existing ones. I got an OK to try to work out the details of a computer programming method and got it to work quite successfully. The method is now known as the Panel Method and is universally used for low speed flow analysis both in this country and in Europe. (Water flows can be analyzed just as well as air flows.) Our first successful calculations were in early 1955. The method is written up in the book *Applied Computational Aerodynamics*, Vol.125 of Progress in Astronautics & Aeronautics, edited by P. A. Henne. Before this method, one could not even calculate the flow about a round body with a hemispherical nose. Afterwards such calculations came easy. Stuart Crump handled research contracts for the David Taylor Model Basin of the Navy. So, once when I saw him, he said the contract he had given us was the best one he had ever given out. For this work I received the Engineering Achievement Award of Douglas Alrcraft in 1958.

At about this same time I had been working on the problem of predicting the transition from laminar to turbulent flow. These are two quite different types of flow that exist, the laminar having much lower drag. At this time, in the early '50's, the methods of predicting this transition point, as on an airfoil, were very poor and quite primitive. They were not based on any kind of sound theory—just empirical. There did exist a theory about the stability of laminar flow, and I thought I might have some success if I developed a method based on it. We worked for several years correlating test data, which was very scarce. In most early airfoil testing, nobody ever determined how much of the airfoil had laminar flow and how much was turbulent; so this work was in an unexplored field. Our studies, while not very satisfying to many theoreticians, did seem to give much better predictions than the other available methods. I proposed a paper about it to the IX International Congress of Applied Mechanics in Brussels. It was accepted and I gave it in 1956. Because on conventional aircraft transition is not a problem, the flow being almost entirely turbulent, my paper received little attention. But one time in the early '60s I was invited to an AIAA dinner for Sir M. J. Lighthill, probably England's top aerodynamic theoretician. He is not much of a talker, but greeted me with "How is your e^9 method doing?" That is the name of my method. I told him all right as far as I knew. He was one of the audience that listened to me in Brussels. In recent years the question of transition became more important as airfoil design methods improved and as smoother surfaces could be made. My theory is not exact, but no one has been able to improve upon it; and now it is extensively used in all airfoil design work. It is now called the e^9 method. The method handles both the two-dimensional flows, like airfoils, as well as bodies of revolution. Now I understand it is being used fairly successfully for three-dimensional flows, like on swept-back wings. It may not be perfect, but there is no substitute. NASA is an extensive user. The Navy found out about our ability to predict transition on bodies of revolution from a call from Rand who was involved in reducing the drag of torpedoes. I got a call one day asking if we knew anything about predicting transition on bodies of revolution. I said we could do that and ended up with a consulting job with them. Later, because this specialty was not a primary concern of the Douglas Co., I proposed to the Navy that we put our potential flow calculations and transition calculations in one package for them and give it to them so Douglas would not be involved in extensive but routine calculations. They bought the idea and we developed the TAPS program (Transition Analysis Program System). The last I heard, it had been sent to the various Navy laboratories and had considerable use. This pro-

AMO and Dincer Cebeci in Istanbul, Turkey, 1973.

Party celebrating the publication of "Analysis of Turbulent Boundary Layers," book by T. Cebeci and AMO Smith in 1974.

Amo and Betty near Izmir, Turkey, 1973

gram was developed at Long Beach.

After developing the Panel Method, I tried to find a general method of solving the laminar boundary layer equations. After considerable study we did find a workable and exact method. Our first report on this was in 1961 from El Segundo.

AT LONG BEACH

In 1961 Douglas closed down the El Segundo plant and consolidated us with the Santa Monica people, all at Long Beach, which was a superior facility and owned entirely by Douglas. Our first laminar boundary layer work, which was under Navy contract, being successful led to several extensions, such as for high speed flow. All these more involved extensions were successful too. As I have said, turbulent flow is usually the more important flow. In laminar flows one uses the regular viscosity of the gas or liquid.

In turbulent flow the regular viscosity does not apply, but a considerable amount of knowledge is available from experiment about what is called eddy viscosity. It was not too difficult to adopt our laminar equations to handle this different viscosity law; so we did it. The accuracy of the calculations, while not as good as for laminar flows, was satisfying. I exposed Dr. Cebeci to this work, and he enthusiastically helped in the development of the method, making all kinds of improvements to the original version of the method. In this turbulent work, since the viscosity used in the equations is an empirical function, there are several different methods that exist, but ours still is holding its own.

Some time about 1970, I received a letter from ACADEMIC PRESS asking me to write a book about boundary layers. I talked this over with Cebeci and proposed a book on turbulent boundary layers, which they bought. Our book was ANALYSIS OF TURBULENT BOUNDARY LAYERS, published by Academic Press in 1974. I supposed it would become obsolete rather fast, but here in 1994 it is still selling fairly well. Cebeci and I together get about $500 a year royalty from it. It seems to sell better in Europe than in the U.S.

To calculate the flow past an airfoil one needs first to calculate the pressure distribution as by the PANEL METHOD and then calculate the boundary layer that flows subject to this pressure distribution. For a meeting in Ottawa I wrote a paper about combining these two methods. Later in 1971 the Canadian Aeronautics and Space Institute gave me a medal for this paper, calling it the best paper at the meeting. The medal is called the Casey Baldwin Award. He was the first Canadian to fly, similar to the Wright Brothers in this country. A general writeup of our work on boundary layers is contained in my paper *The Boundary Layer and I*, AIAA journal, Vol.19, No. 11, November 1981.

In 1969 I was elected a fellow of the AIAA.

I will close by mentioning one more technical subject. Late in 1971 I was asked by the AIAA to give a talk on "high lift aerodynamics" to be given in August 1974. I accepted and started to work on this subject. There was a lot of ignorance and confusion in this field, such as understanding how a slotted flap or a slat worked. So, it was a good subject to talk about. I did a lot of study and made extensive calculations. I gave the paper, which was called the Wright Bros. Lecture at an AIAA meeting in August 1974. The paper seemed to fill a void in the knowledge about how to design airfoils for developing high lift. The AIAA paid for me to repeat the lecture around the country. I repeated it 19 times in the U.S. and the 20th time was in Shanghai. Prof. Eugene Covert of MIT in 1989 told me they were still prescribing it as required reading in one of their aerodynamics courses at MIT. Also, a few years back I received a letter from a Boeing engineer saying that my paper was prescribed reading for any-

AMO and Betty at the Palace Museum, Beijing, China, 1979.

one entering into the wing design field there. This engineer was giving a paper in Brussels in 1984 entitled, more or less, "ten years of development since AMO Smith's paper on High Lift Aerodynamics".

MISCELLANEOUS

In 1975 I was given an honorary D.Sc. degree from the University of Colorado. As far as I know, one of the U. of Colorado professors contacted a Cal Tech professor about suggestions to their Board of Regents. The Cal Tech Professor, Milton Clauser, suggested me.

After retiring, in 1979 I received an invitation to give a series of lectures in China on what we had done in aerodynamics, to be given at the Beijing Institute of Aeronautics and Astronautics. In the original letter they said they would pay part of my wife's expenses, but not all. I wrote back accepting, but asked what she would do all the time I was lecturing. They wrote back that they needed someone to help teach English, or more precisely "American". Betty said she would do that, so we had a good deal. All our expenses were paid, plus/60 yuan a month for spending money (a yuan was worth about 65¢) and two weeks vacation touring China at the end of the lecture period. In the original letter they suggested a two or three month visit; so I compromised and we stayed 10 weeks. I had a Purdue graduate translator all the time. Since my students did not speak English we had very little interaction. However, Betty's English class was at a 3000 word level, and there was rather good communication. What they really wanted in this class was not to learn from Betty more grammar, etc. but to learn "American" and about America. Their textbooks were all from England. Our visit was very enjoyable. I would give lectures in the morning and then be free all afternoon. I had prepared most of my lectures in the U.S. to be sure I had all the material I needed. Thus, most of the afternoons and evenings were quite free to shop, sightsee, etc. The English class would assign one of the students to be our guide for the day. The last two weeks we vacationed first to Hangchow, then Guilin, then Canton, and finally Shanghai.

In Shanghai there was an Aeronautics Society that wanted me to give a lecture. I said I could talk on a number of different subjects. They chose my High Lift Aerodynamics subject, making this my 20th lecture on the subject. I went to China in September 1979 and returned late in November. Betty followed me a few weeks later; and in the meantime my translator and I flew to Shenyang in Manchuria to visit an aircraft factory.

I consulted for about 10 years from 1975 to 1986 for MDAC (McDonnell Douglas Astronautics Corp.). A torpedo is a kind of guided missile in water; and these rocket guys hoped they could get some business in this area. From my contact with the Navy, I knew something about water flows, which the rocket people didn't. So, they hired me to help out. Since they could never break into the field, they laid me off. I also had a number of other consulting jobs.

In 1985 I received the Fluids Engineering Award of the ASME (American Society of Mechanical Engineers). In 1986 I was elected an Honorary Fellow of the AIAA. Allen Puckett, then President of Hughes Aircraft, was the other American. Three are elected, the third being foreign. In 1989 I was elected to the NAE (National Academy of Engineering) and in 1990 was elected into Who's Who in America.

AMO's election to the National Academy of Engineering in 1989

AMO and his National Academy of Engineering Colleagues

AMO and Betty, 1986

AMO and Betty in 1987

AMO and his family in 1993

Examples of AMO's Craftsmanship

A Rocking Horse

A Hanging Cabinet

Bird Houses

A Cherry Hexagonal Table

A Curio Cabinet

Cut-out Fence at AMO's Mountain Home

Ladder Chair

A Doll Bed

A Gossamer Condor Whirligig

A Dog Cart

AMO's wind tunnel

3

Laminar Flow Control and Stability/Transition

by
Kalle Kaups
A member of AMO's Research Group

It is relatively easy to see that many of the areas to which AMO contributed originally were related to laminar flow control work. The best source for this is the article "Boundary Layers and I" written by AMO in his retirement years, it is highly readable and highly recommended.[64] Since this is the case, one is not tempted to improve on it except to try to fill in with additional tidbits of information culled from reports, papers still in existence and from memory.

His becoming involved in laminar flow control problems was more complicated than his Master's thesis at CALTECH in 1938, "A Preliminary Study of the Laminar Flow Control," would indicate. According to his own writings, the topic dealt with boundary layer control applied to maximum lift enhancement, and that his main interest was in aircraft preliminary design. Then followed the World War II years with more urgent distractions but the laminar flow control problem was re-examined starting in 1944, possibly in conjunction with the then novel jet propulsion. A paper on the same subject was published in the JAS in 1947[2], and discussed the use of boundary layer air for propulsion, including distributed suction to keep the flow laminar on portions

*References are at the end of the book

of the wing, fuselage and tail surfaces. This paper led directly to a US Navy contract to study the problem further in 1948/1949 and resulted in three company internal reports dealing with turbojet engines utilizing boundary layer air, engine powered pumping system, and tests of flow characteristics through porous and perforated sheet materials. The origin of the contract to design a laminar flow control airfoil is now even more obscure. The only illuminating statement is found in an unnumbered company review report dated July 1953 as follows: "Contract was obtained in 1947 to design a slotted laminar flow control airfoil based on Pfenninger's work in Switzerland but for higher Reynolds numbers." By his own admission, AMO considered this multi-slot arrangement as too complicated and he was looking for a simpler solution. This occurred to him after listening to Sidney Goldstein's 11th Wright Brothers Lecture in December 1947. He proposed using few suction slots distributed along the chord instead of a single slot per surface near the trailing edge. The resulting airfoil with a mildly sawtooth pressure distribution was expected to have almost regular characteristics with the suction system inoperative. The design was a success beyond expectation: at the limiting tunnel Reynolds number of $R_c = 4.25 \times 10^6$ laminar flow was maintained to 98 percent chord versus the predicted 90 percent. This result is quite astounding if one considers the design tools available at the time. Aside from the relatively sophisticated inverse potential flow method used, the final stability analysis was based on the measured boundary-layer profiles and not on calculation! The stability analysis itself was rudimentary as well: the critical Reynolds number for Tollmien-Schlichting waves was determined from Lin's short method which required second derivatives of the velocity profiles, and the critical Reynolds number for Görtler stability was fixed by Görtler number equal to 9 as determined by Liepman's experiments.

Whereas the first design, a symmetrical section at zero lift, was used to demonstrate the feasibllity, the follow-on design was for a realistic high speed airfoil which sights set at obtaining fully laminar flow at chord Reynolds numbers of around 50×10^6, a very ambitious goal. However, before the design could proceed, the tools had to be updated and AMO was "diverted" to analytical work. In a short order several improvements appeared. The capability to calculate Falkner-Skan profiles was in place by 1951, and one suspects that the eventual publication of references (4, 11) resulted from this work. For stability analysis AMO was aware that he needed to calculate the growth of amplifications instead of the critical Reynolds number. This need was fulfilled in the case of Tollmien-Schlichting waves by Pretch's charts, which were adopted but required a fair amount of interpolation and hand calculations. The really original work was done in solving the Görtler stability problem with complete set of amplification curves. Most of this work was finished by the spring of 1952 and subsequently found publication in[7]. This time frame may be considered as the birth date of the e^n-method. AMO found that the available data on Görtler vortices correlated well with a e^{10} but data on Tollmien-Schlichting waves had considerable scatter so that he expressed some doubts about its validity. However, as additional data became available and was analyzed, the original assumption was proven beyond doubt[9]. The final design, called DESA-2, was a 6-foot chord cambered 6.6 percent thick airfoil with 8 suction slots on both the upper and lower surfaces. The preliminary tests were conducted at GALCIT where at $R_c = 6.5 \times 10^6$ some loss of stability was experienced which was assigned to mismatch in suction slots. This was subsequently corrected and the model was shipped to Langley for high Reynolds number tests. To everyone's surprise the identical problem reappeared and no amount of fine tuning cured it. Despite hot wire measurements, no obvious reason presented itself and the following suspects remained: slot design (step-down), roughness and turbulence. It was later explained that Görtler vortices do not damp out when they enter a convex surface region, they remain dormant and assume their growth over the next concave surface region.

This was essentially the end (July 1953) of this phase of the investigation that started out with high hopes and very ambitious goals. By one account, the US Navy had funded approximately 46,700 man-hours worth on laminar flow control work during the seven years between 1949 and 1954, a clear measure of the effort involved! Indication that not all was wasted and that Navy confidence remained, additional contracts followed, but now directed toward practical problems, at least on the surface because there was considerable leaveway as to what problem was to be attacked. The best known results from this contract series was related to experimental work on the smallest roughness that causes transition[13,14]. This may have been prompted by the desire to get to the bottom of the earlier problem. Another interesting company internal report from this time was titled "Studies and Experiments in Drag Reduction by Laminar Flow Control."[9] Tests included artificial thickening of the boundary layer near the nose of an airfoil to reduce boundary-layer sensitivity to roughness, attempted to use resonance to damp boundary-layer oscillations; porous slots for suction; combined blowing and suction on an airfoil to reduce turbulent skin-friction; and longitudinal groves to make the boundary-layer more three dimensional and reduce skin-friction. This idea sounds very close to the Vgrooves used today expect for the larger dimensions employed.

From the end of this period, about 1957, AMO concentrated more and more on the development of potential flow and boundary-layer calculation methods which are discussed elsewhere. However, he retained his interest in laminar flow control and stability. He was still intrigued by the range of validity of the e^n-method. When Pfenninger's experiments in a round tube became available, he re-examined the whole transition problem for this configuration and extracted from the entrance flow results that e^n-method should be valid at least to $R_c = 46 \times 10^6$ [15]. Appreciation for his expertise in this area can be gauged by the invitation to serve on USAF advisory committee for the X 21 program. I remember one conversation with him regarding the conditions under which laminar flow control would become feasible, if at all. His unhesitated answer was yes, but only if the design is driven by aerodynamic consideration alone, any compromise is bound to lead to failure. In another instance, showing AMO's practical side, when asked how does one know when a surface is rough? His answer: "use your touch, if the surface feels rough, it will be rough, laminar flow cannot exist and there is drag penalty in turbulent flow to boot"!

In the following years his time was increasingly taken up by supervising tasks because his group had grown in numbers and because everybody was asking his advice and guidance. There was nearly a ten year period when boundary layer stability research was on the low burner partially because of other pressing problems and because AMO saw it as a taunting task to do anything meaningful computationally in an economic way. But as computer capability increased, around mid sixties, AMO became more optimistic about the possibility of solving the stability problem for high Reynolds numbers. When Dr. Wazzan from UCLA became consultant to the group, AMO encouraged him to look into the problem. As it happened, at that time, Landahl and Kaplan had come out with a purification scheme for the numerical solution of the stability equation which allowed a considerable extension of the solutions into the high Reynolds number range. Additionally, the addition of Gram-Schmidt orthogonalization procedure removed all the restrictions on Reynolds number. The first demonstration of this new method was applied to Falkner-Skan separating profile[32]. Almost simultaneously at AMO's suggestion, investigation was conducted on the effect of heating and cooling on water boundary layer stability as indicated by viscosity variation with temperature: heating should enhance stability in water, which is opposite to the heating effect in gases. As a matter of fact, this analysis[33] showed that substantial increase in the critical Reynolds number occurs with relatively small wall temperature increase. This opened up another avenue to obtain laminar flow control in a less complicated way although in another environment. Further investigations covered the effect of pressure gradient and showed that for a given freestream temperature, there is a wall temperature at which the critical Reynolds number is maximum[46,49]. The culmination of this research produced a fully automated design program for the US Navy that combined potential flow, boundary layer and stability calculations to predict the extent of laminar flow obtainable by the e^n-method. This makes one wonder whether he ever dared to anticipate the end product when he started his investigation practically from the scratch? It must have been gratifying indeed!

Since AMO also was a good housekeeper who liked to do some cleanup work, he initiated the recalculation of the Pretch's stability charts and a check on the e^n-method[41]. To his satisfaction all previously calculated results based on approximate methods were substantiated by the more exact calculations, now making use of spatial amplification theory.

Another advantage of the e^n-method that surfaced during these investigations was that when other investigations applied the standard R_θ - R_x correlations to transition prediction on bodies of revolution either transition could not be predicted, or that in the inverse solution cases resulted in bodies which could not sustain laminar flow. A US Navy sponsored research project established that the e^n-method was the only feasible way to predict transition on bodies of revolution at that time. The basic fact here was that profile shape in the form of the parameter H is more important than its momentum thickness. Further elaboration of this aspect of transition prediction in general is found in a note[62] on H–R_x correlation to which AMO was a co-author.

After AMO retired from the Douglas Aircraft Company, he joined the Mechanical Engineering Department at UCLA as Adjunct Professor, acting in an advisory capacity for work on the stability of water boundary layers and transition under US Navy contracts. His contributions appear in publications under the previously mentioned references 61 and 62, which are a summary of his accomplishments. This work was completed about 1980, but this did not indicate the end of his association with water flows as seen from a reference dated 1987[69]. Additionally, he kept active in his retirement years by considering the application of laminar flow control to sail-planes in two papers[66,67]. Besides formal papers and reports, his contributions appeared many times in lectures, workshops and discussions. At the end of this book is a list of the oral presentations as well as they are known today, but there is no guarantee for completeness because some of them were classified and never mentioned in public. It may be fitting that his activities in aeronautics had come to a full circle; they started with building and flying glider in his high school days and ended with outlining laminar flow control prospects for sail-planes.

Transition, Pressure Gradient, and Stability Theory

A. M. O. SMITH
Supervisor, Aerodynamic Research
Douglas Aircraft Company, Inc.
El Segundo Division
Presented at
IX INTERNATIONAL CONGRESS FOR APPLIED MECHANICS
Brussels, 1956

Abstract

This article presents a method of predicting transition that is founded upon the theory of boundary-layer stability. It considers both plane and rotationally-symmetrical flows of a very low-turbulence incompressible stream past very smooth bodies.

The theory and its implications are briefly reviewed and discussed. Nearly all the applicable experimental transition data have been studied, and correlated with calculations made with the aid of Pretsch's stability charts. A strong correlation of experimental results with this theory was discovered. In fact, to first order, Tollmien-Schlicting waves are found to undergo an apparent, amplification ratio, exp $\int \beta_i dt$, equal to about e^9 by the time transition begins. This value was then used to predict transition for a considerable varity of flows, both in wind tunnels and in flight. The predictions agree quite well with experiment, particularly in view of the nature of the experimental data.

In contrast, R_{δ^*} or R_q at the beginning of transition vary greatly with pressure gradient and are not at all suitable alone as fundamental guides for locating transition. As a measure of transition on bodies of revolution R_{δ^*} is even less satisfactory than it is for two-dimensional flows. However the method based on stability theory correlates both types of flow with equal accuracy.

A short-cut empirical method of predicting transition, due to Michel, has been studied. In two-dimensional flow its validity was found to be good, and the reason for its success can be explained by stability theory.

1. Introduction

To the best of the writer's knowledge, three methods for estimating the transition point are currently in existence, namely:

I. The use of an allowable value of R_{δ^*} or R_θ as a criterion; when some arbitrary value is exceeded, transition is expected to occur.

II. The use of the neutral-stability limit supplied by boundary-layer stability theory.

III. Michel's method [1], which uses an empirical relation between $R_{\theta_{tr}}$ and $R_{x_{tr}}$

The first has no foundation in theory other than the fact that a boundary layer becomes turbulent when its Reynolds number becomes large enough. In particular, no method is available for changing the value of R_{δ^*tr} due to variations in pressure distribution. R_{δ^*tr} is the Reynolds number at transition based on displacement thickness.

The second does have its foundations in a physical process, the growth or decay of Tollmien-Schlichting waves in the boundary layer; but the method is of only qualitative value, because the large interval between the neutral-stability point and the transition point is left unexplained.

The third, Michel's method, overcomes the most glaring weakness of the first, because it does implicitly allow for the shape of the pressure distribution in determining a critical value of $R_{\theta_{tr}}$, the Reynolds number at transition based on momentum thickness. But the method is strictly empirical, and because no reason for the successful correlation has ever been advanced, the method has been received with considerable reserve.

2. An Extended Method Founded On Stability Theory

Description:

Method II does have a solid theoretical foundation based on a physically conceivable process, but it leaves a large void between the neutral point and the transition point. One method of filling that void is to continue the same analysis past the neutral point by calculating the growth of Tollmien-Schlichting waves up to the very point of transition. Then if the theory is sound, all effects of pressure distribution will be properly taken into account and even though the transition process itself is not understood, it may be possible to arrive at a single correlation constant. This idea has been known for some time but has never been carefully studied. One drawback, of course, is that stability theory is a small-disturbance type, whereas it is well known that the disturbances in the boundary-layer become large shortly before transition.

Liepmann* has perhaps enunciated the most logical basis for correlation of wave growth with the transition point. He makes the following statement about $R_{x_{tr}}$, the x-Reynolds number for transition:

" $R_{x_{tr}}$ is the Reynolds number at which the apparent shear $\tau_A = -\rho\overline{u'v'}$ due to amplified boundary-layer oscillations at any point in the boundary layer becomes equal to the laminar shear $\tau_L = \mu\, d\bar{u}/dy$ in the boundary layer."

Liepmann derives an approximate formula for application of the above principle that incorporates a quantity calculable by stability theory namely the amplification ratio $a(R_{\delta*})$. The formula is:

$$\frac{\tau_{A_{max}}}{\tau_L} \approx -\frac{2}{c_{f_L}} \left\{ kb \left(\frac{u'}{U_n}\right)^2 \left[a\left(R_{\delta*}\right)\right]^2 \right\}_{max} \tag{1}$$

When $\tau_{A_{max}}/\tau_L \cong 1$, transition should occur. Hence, Eq. (1) states that transition will be determined by the local skin-friction coefficient, the turbulence structure of the stream, and the amplification factor.

The Pretsch Charts:

Eq. (1) makes no assumption that the disturbance is a Tollmien-Schlichting wave but if it is, it becomes possible to compute the amplification factor by means of Pretsch's charts.[3] Pretsch computed the amplification rate of such waves for several of the two-dimensional wedge flows, namely those with Hartree β = -0.198: -0.10, 0. 0.2. 0.6. and 1.0. Two types were computed; rate charts showing the rate of growth of disturbances of various frequencies and cumulative amplification charts for constant-β flows, showing the total amplification up to any value of $R_{\delta*}$. Pretsch has shown that for boundary-layer flows in a stream of non-constant velocity, the basic analysis is valid whether they be two-dimensional or axially symmetric. If the frequency β_r of a certain wave, the boundary-layer Reynolds number $R_{\delta*}$ and β are known, one can read from his charts the amplification-rate coefficient β_i of this frequency. Therefore, during a time interval (t_1, t_2) a particular frequency will undergo an amplification of amount:

* For further details, documentation, etc., see Ref. 2.

$$a(R_{\delta *}) = \exp \int_{t_1}^{t_2} \beta_i \, dt \qquad (2)$$

By now the procedure indicated by Eq. (1) should be clear. In determining the value of $\tau_{A\,max}/\tau_L$ at some point on a surface, one seeks the frequency that will produce the highest value of the subject quantity. The highest value of the product $(u'/U)_n\,[a(R_{\delta *})]$ is being sought, not the highest value of $a(R_{\delta *})$ alone. Therefore, a careful evaluation of Eq. (1) demands at least a frequency spectrum to supply data on initial values of the disturbance. A mean value of u'/U is unsatisfactory, for the highest value of $a(R_{\delta *})$ may occur at some frequency for which u'/U is low, but at some other frequency u'/U may be enough greater to produce a larger value of the product $(u'/U)_n\,[a(R_{\delta *})]$ even though $a(R_{\delta *})$ is below the maximum. In short, the product should be maximized.

But, as will be seen later, the amplification factor is of the order of 8000 and by comparison, variations of c_{f_L} and the turbulence term in Eq. (1) are so small that they can be neglected with respect to $a(R_{\delta *})$ in any first-order analysis.

Method For Correlation With Experimental Data:

Assuming the flow is not of a constant-b type, the ideal procedure for analyzing the experimental results would require: (a) accurate calculation of boundary-layer profiles along the entire given flow; (b) for these specific profiles, calculation of rate charts similar to those of Pretsch; (c) use of these charts for computation of the highest amplification factor.

Since such would be a colossal undertaking, growth of disturbances was computed for the study described here in a much cruder fashion designed to use Pretsch's charts. His charts require knowledge of $R_{\delta *}$ and β. These two required properties were computed by the method of [4], a rapid method that uses β as a parameter and has a high accuracy for most accelerating flows.

According to stability theory, disturbances will grow as $\exp \int \beta_i \, dt$. Using time as a parameter when dealing with flow along a surface is inconvenient, but it can easily be eliminated by using c_r, the phase velocity of advance of a particular wave along the surface, that is, $dx = c_r\, dt$. By means of this relation it is possible to rewrite $\int \beta_i \, dt$ in terms of several dimensionless quantities that occur in Pretsch's charts:

$$\int_{t_n}^{t_1} \beta_i dt = \int_{(x/c)_n}^{(x/c)_1} \frac{\beta_i \delta*}{U} \left[\left(\frac{U}{U_\infty}\right)^3 \frac{R_c}{R_{\delta*}^2} \right] \frac{\alpha \delta*}{\frac{\beta_r \nu}{U_\infty^2}} d\left(\frac{x}{c}\right) \qquad (3)$$

Eq. (3) can be used to compute the amplification of any frequency. For purposes of correlation, if x_{tr} is given, one seeks the maximum value of Eq. (3) by systematic computation of the growth of waves of several different frequencies. The necessary quantities are read from Pretsch's charts.

A Sample Calculation

A sample calculation as applied to a body of revolution at $R_c = 6.65 \times 10^6$ is illustrated by Figs. 1, 2 and 3. The body of revolution and boundary-layer properties as computed by the method of [4] are shown in Fig. 1. Notice that the boundary layer becomes unstable to Tollmien-Schlichting waves near 6 percent of the chord, whereas, according to Fig. 3 the experimental transition point is near 50 percent of the chord at this Reynolds number emphasizing the criticism of method II.

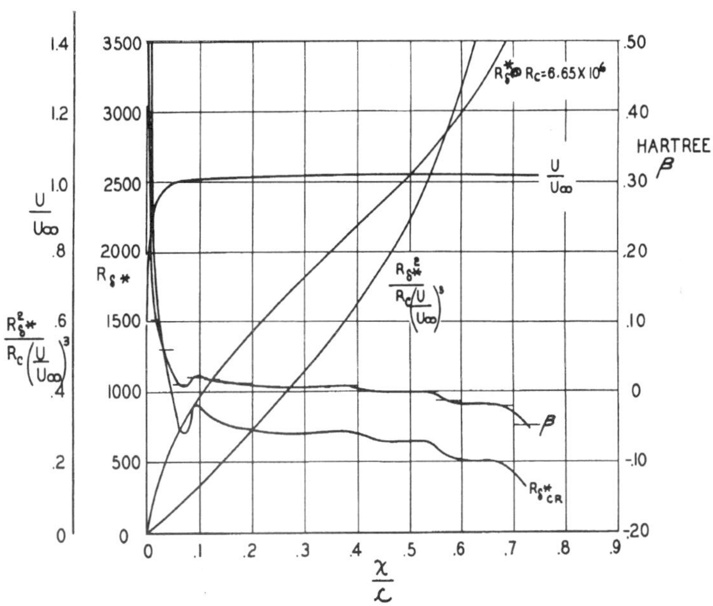

Figure 1 Boundary layer properties, N. A. C. A. body of revolution, of fineness ratio 9.0. x is the distance from the nose measured along the surface.

The correlation problem requires computation of the apparent amplification ratio for the worst frequency at the measured transition point. Fig. 2a is a plot of the integrand of Eq. (3) for several values of $\beta_r \nu / U_\infty^2$ chosen to be near the critical. The highest frequency responds first, but as the boundary layer thickens farther downstream, a lower frequency receives the greatest amplification and the high-frequency wave receives relatively less amplification or may even decay. Furthermore, the amplification rate may grow and decay and then grow again according to the variation of b. Such an event is illustrated by the case $\beta_r \nu / U_\infty^2 = 2.5 \times 10^{-5}$. The decelerating flow behind $x/c = 0.7$ so destabilizes the boundary layer that disturbances of this frequency receive new energy and commence rapid growth again. Since the transition point is known to be at $x/c = 0.50$, one seeks the maximum value of the integral up to this point. The insert in Fig. 2a is an auxillary interpolation plot for determining the maximum value. It illustrates the considerable sensitivity of amplification to frequency.

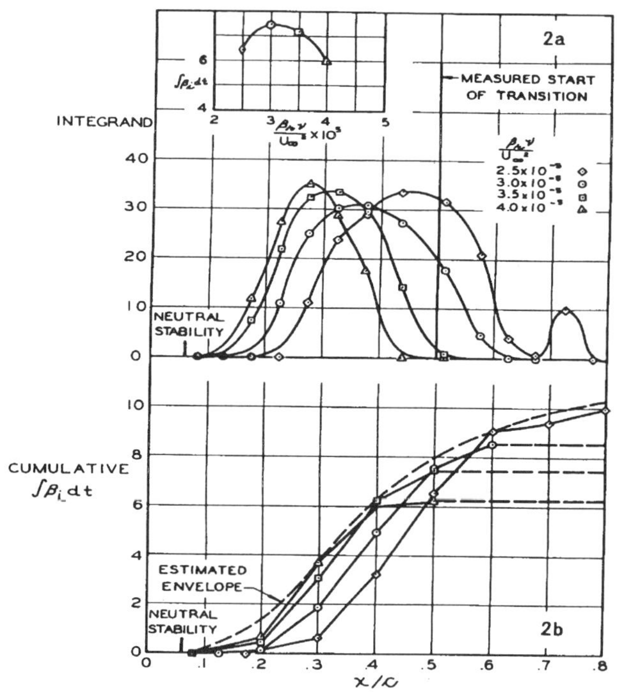

Figure 2. Tollmien-Schlichting wave growth at $R_c = 6.65 \times 10^6$, N. A. C. A. body of revolution of fitness ratio 9.0.

In the converse problem of estimating the transition point, one must determine where the most critical frequency first reaches its limiting value, for example, e^9. Fig. 2b illustrates the procedure necessary now. One must construct the cumulative integral and study enough values of $\beta_r \nu / U_\infty^2$ to deter mine an approximate envelope. The results for a range of Reynolds numbers are shown in Fig. 3 Curves of R_{δ^*} = 3000 and 5000 as well as curves for $\int \beta_i dt$ = 0, 8, and 9 are shown to compare several different transition criteria. Notice that ò b$_i$dt = 8 predicts transition very accurately.

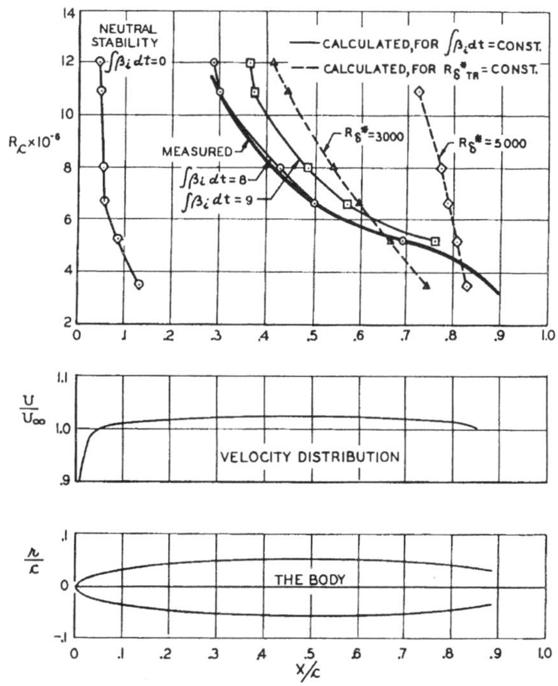

Figure 3. Comparison of two methods for estimating transition, N. A. C. A. body of revolution, fineness ratio 9.0. X is distance from nose measured along axis.

3. Results-Evaluation of $\int \beta_i dt = 9$ As A Transition Parameter

The initial correlation study consisted of computing the maximum apparent amplification ratio at the measured transition point for a large variety of cases. The tests analyzed are virtually all existing tests that are at sufficiently high Reynolds number and in very low-turbulence tunnels or in flight. The types of case covered are:

a. Bodies of revolution in low-turbulence wind tunnels.
b. Airfoils in flight.
c. A convex plate in a narrow low-turbulence channel.
d. Airfoils in low-turbulence wind tunnels.

Forty-eight cases were studied covering 28 different body shapes in 9 test environments one of which was free flight.

In the original study[2] no discernible variation of $\int\beta_i\,dt$ was found, but the scatter was large. Because the parameter $\int\beta_i\,dt$ appeared sensitive, it appeared that the best way of testing the theory was to put it to work, that is to predict the transition points for all the cases studied by means of a single value of the amplification factor. Preliminary examination of the data indicated that "9" was the best integral value for the exponent so that the factor becomes e^9. Fig. 4 presents the results. In view of the nature of the data and the poor methods of calculation, the agreement is surprisingly good. Observe that there is no discernible indication that $\int\beta_i\,dt$ should be other than constant, that this result seems to apply over a range of nearly two cycles, and that it is valid for both airfoils and bodies of revolution.

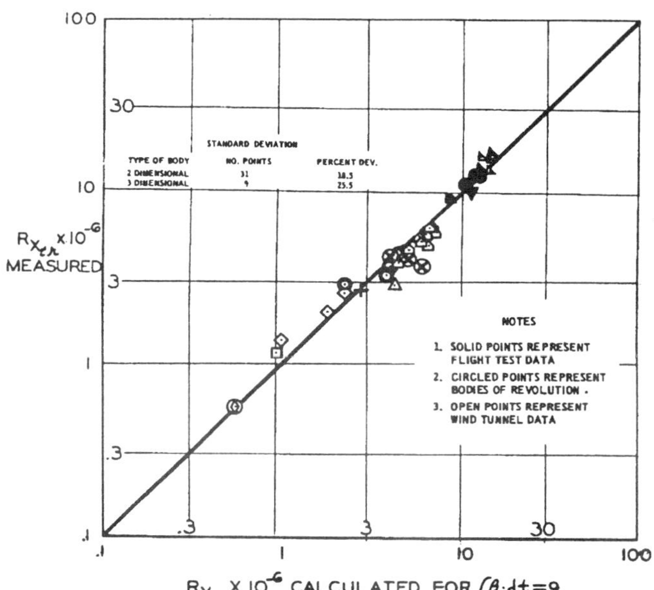

Figure 4 - Transition computation - correlation between measurement and prediction by stability theory.

4. A Short-Cut Method (Michel's Method)

Michel [1] discovered that experimental transition data could be correlated well by means of a plot of $R_{\theta_{tr}}$ versus $R_{x_{tr}}$, so that estimation of transition became very easy, requiring only static-boundary-layer calculations. However, since the reason for the successful correlation remained unexplained, the method was not used with confidence. It is shown in [2] that the locus of $\int \beta_i dt = 9$ is identical to Michel's curve, within experimental accuracy. Hence, Michel's correlation is explained by stability theory and does have a sound underlying physical concept. Fig. 5 presents all the experimental transition data in the ($R_{\theta_{tr}}$, $R_{x_{tr}}$) coordinate system. Values for bodies of revolution are plotted in terms of Mangler's transformed values. Shown also are Michel's original line, the line $\int \beta_i \, dt = 9$ for similar flows, and a proposed new fit to the data. The new one is proposed for the reason that the present data cover a wider range of Reynolds numbers than Michel's. Because the method is not exact, it is acceptable to represent the data by means of a straight line on a logarithmic plot. The equation:

$$R_{\theta_{tr}} = 1.174 \quad R_{x_{tr}}^{0.46} \qquad 0.3 \times 10^6 < R_{x_{tr}} < 20 \times 10^6 \tag{4}$$

is proposed as a criterion.

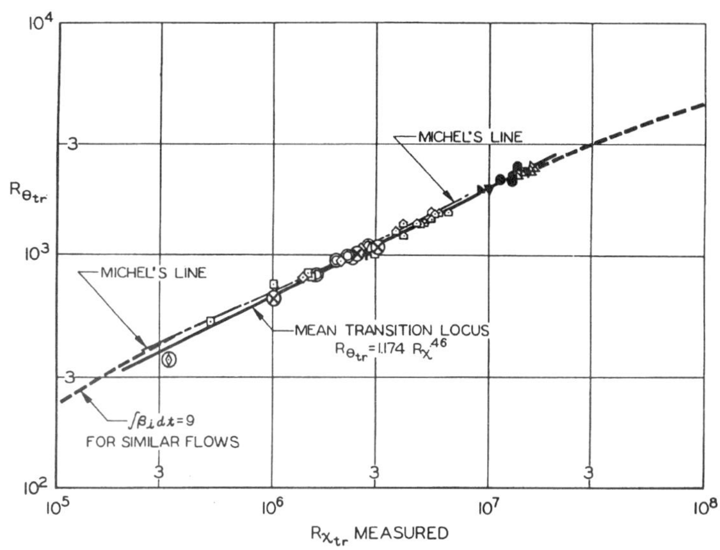

Figure 5 Summary of experimental transition data, $R_{\theta_{tr}}$ - $R_{x_{tr}}$ plot.

It has been used to predict transition for all the cases considered and results are compared with the test data in Fig. 6. For two-dimensional bodies, the method is equal to the direct method of stability theory, but for axially symmetric bodies it is significantly inferior. Therefore, in two-dimensional flows the method of Michel can be used with confidence, provided the pressure distribution under consideration roughly resembles one or more of those used in the original correlation. However, the correlation can hardly be called satisfactory for axially symmetric flows.

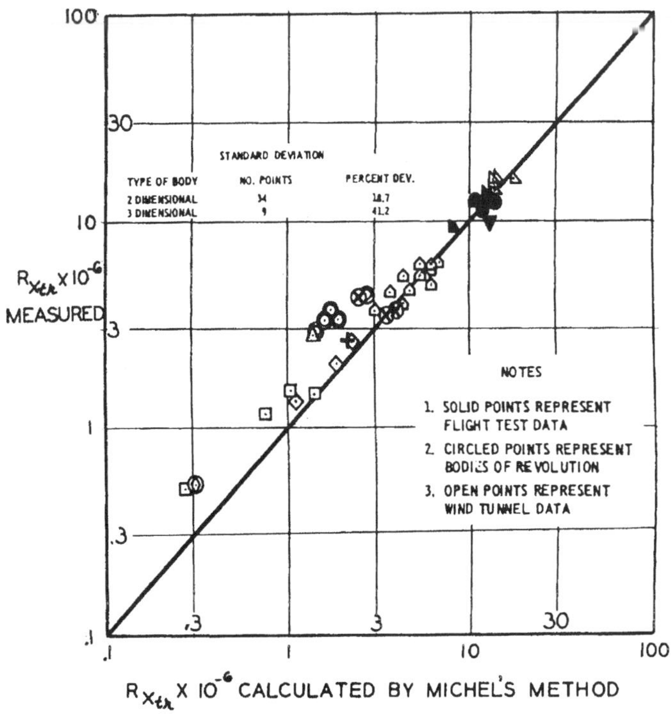

Figure 6. Transition computation - correlation between measurement and prediction by Michel's method, using Eq. (4).

5. Conclusions

1. Stability theory successfully explains the effect of pressure gradient on the transition point in low-speed flows of very low-turbulence streams past very smooth bodies, either rotationally-symmetrical or two-dimensional.

2. By the time the transition point is reached, the most critical Tollmien-Schlichting wave has undergone an apparent amplification of about e^9 times its amplitude at the beginning of its growth. This particular value has been found by means of Pretsch's charts. It might change if the charts were computed more accurately.

3. By means of the constant, e^9, it is possible to predict the transition point in arbitrary flows with an error in $R_{x_{tr}}$ seldom greater than 20 percent.

4. Evidently, in view of the above successful correlation, the unstable, self-energizing mechanism of Tollmien-Schlichting waves is the dominating process during the initial stages of the evolution of a boundary layer from a laminar to a turbulent state when the flow meets the limitations of this report.

5. The considerable body of data examined indicates that the extent of laminar flow obtainable in good low-turbulence tunnels is no less than that obtainable in flight.

6. The experimental data demonstrate that R_{δ^*} and R_θ do not have constant values at the transition point. Rather, $R_{\theta_{tr}}$ and R_{δ^*} are variable but can be correlated approximately by Michel's method.

7. Stability theory supplies the explanation for Michel's hitherto unexplained transition-correlation curve.

8. Michel's correlation method is extended by means of Mangler's transformation to include rotationally-symmetric flows, but the results have a low degree of accuracy.

9. The data and studies of this report justify Michel's method as a rapid, approximate method for estimating the transition point in two dimensional flow, so long as the velocity distribution of the problem roughly resembles one of the distributions used in the original correlation.

6. References

1. Michel, B., Etude *de la Transition* sur les *Profils* DAile-Etablissement d'un Point *de Transition et Calcul de laTrainee de Profil en Incompressible*, ONERA, Rapport 1/1578A, July, 1951.
2. Smith, A. M. 0. and Gamberoni, Nathalie, *Transition, Pressure Gradient and Stability Theory*, Douglas Aircraft Co., Rept. No. ES 26388, Aug. 31, 1956.
3. Pretsch, J., *Calculation of the Limit of Stability of Boundary Layer Profiles* and of the *Amplification of* Disturbances. A. V. A. Monographs, British Reports and Translations No. 1004, March 15,1948.
4. Smith, A. M. 0., *Rapid-Laminar Boundary Layer Calculations* by *Piecewise Application of Similar Solutions*, accepted for publication in the Journal of the Aeronautical Sciences.

Bob Liebeck, Andy Bauer and John Hess with AMO.

4

Potential Flows

by
John L. Hess
A Member of AMO's Research Group

In order to appreciate the magnitude of AMO Smith's contribution in the area of calculation of potential flows, it is necessary to be aware of the situation that existed before he turned his attention to the problem Specifically, the desired solution is that of the direct problem of potential flow where a body shape is given and the resulting flow field is required along with velocity and pressure distributions on the surface of the body. When AMO began his work, methods of solving this problem were limited to two-dimensional and axisymmetric slender bodies having small perturbation velocities, i.e., total velocities close to that of the free stream. (There were also special solutions, such as ellipsoids and, in two-dimensions, conformal mapping solutions, but these last required them mapping of the body to a circle, at that time an unsolved problem.) AMO solved this problem by applying his special genius for combing the most abstruse theoretical results with engineering insight. He formulated the problem as an integral equation over the body surface for a source distribution. After his initial formulation, he collaborated with Jesse Pierce to develop a fully working computer program. The results are reported in Ref. 12 for non-lifting two-dimensional flows and axisymmetric flow. Ref. 12 thus contains the first general numerical method for bodies of arbitrary shape and perturbation velocities of arbitrary size - a real breakthrough.

*References are listed at the end of the book.

Moreover, surface pressure distributions (the main interest) may be obtained without having to calculate the entire flow field - a huge efficiency. Shortly after Ref. 2 was published, Jesse Pierce retired and I took over the work with the two goals of extending it to axisymmetric bodies at angle of attack and two-dimensional lifting flows.

Even before this work was completed, AMO had approached the Navy and obtained a contract for us to generalize the work to an arbitrary three-dimensional body. At the time (1959) this was unprecedented in two ways. First, nobody even thought of calculating general non-small-perturbation three-dimensional flows. Second, we in industry didn't think in terms of research contracts (hard as that is to believe now). Soon AMO entrusted me to carry out this work. One of the key features of the method was the approximation of the three-dimensional body surface by a large number of small plane quadrilaterals. The appearance of such an approximate body led to the technique being named a "Panel Method," although not initially by us. Upon successful completion of this work in 1962, a report was written which finally led to Ref. 20. This reference is thus the seminal paper on which all subsequent panel-method work, by ourselves and others, is based.

In 1966, AMO and I authored a major review article (Ref. 28) summarizing the potential flow work accomplished over the past ten years by ourselves and Joe Giesing - a remarkable decade of work initiated and supervised by a remarkable scientist—AMO Smith.

EXACT SOLUTION OF THE NEUMANN PROBLEM. CALCULATION OF NON-CIRCULATORY PLANE AND AXIALLY SYMMETRIC FLOWS ABOUT OR WITHIN ARBITRARY BOUNDARIES

by
A. M. O. Smith and Jesse Pierce
Douglas Aircraft Company

Abstract

An exact general method of solving the Neumann or Second Boundary Value Problem has been developed and has been applied to the calculation of low speed non-circulatory flows about or within bodies of almost any shape, provided the flow is either plane or axially symmetric. Solid-body inlet and purely internal flow problems can be solved. Because the method is capable of dealing with several bodies at once in the presence of one another, problems of interference can be treated with ease. Boundaries need not be solid, that is, flows involving area suction can be calculated. With an IBM 704, calculations require from 3 minutes to 2 hours according to the complexity of the body.

1. Introduction

The general subject of this paper is exact solution of the Neumann problem for arbitrary boundaries, with application to non-circulatory plane or axially symmetric fluid flows, either internal or external. Incompressible inviscid fluid-flow problems are problems of the Neumann type because the field is harmonic and velocities $\partial\varphi/\partial n$ normal to the boundary, at the boundary, are specified. As is shown in any good treatise on potential theory (e.g. [1]), exact solutions of the Neumann problem can be obtained from the following integral equation for s, the surface density.

$$\frac{\partial\varphi}{\partial n_-}(p) = 2\pi\sigma(p) + \iint_S \sigma(q)\frac{\partial}{\partial n}\left(\frac{1}{r}\right) dS \text{ (Region R)} \tag{1a}$$

$$\frac{\partial\varphi}{\partial n+}(p) = -2\pi\sigma(p) + \iint_S \sigma(q)\frac{\partial}{\partial n}\left(\frac{1}{r}\right) dS \text{ (Region } R'\text{)} \tag{1b}$$

The notation and measuring system are indicated in Fig. 1 and S is the entire surface area. For problems of fluid flow, (1) may be thought of as a formulation in terms of a surface distribution of sources and sinks whose strength at any point is $\sigma(q)$. Equation (1) is a Fredholm integral equation of the second kind. It is solved under the restrictions already mentioned for either the region R or R', Fig. (1).

The general procedure used in the analysis is as follows. The body profile is specified by an arbitrary number of coordinates, not necessarily equally spaced. These coordinates serve to divide the body into elements as indicated in Fig. 1. The boundary condition at any point p supplies values of $\partial\varphi/\partial n$ at that point, that is, values needed for the left hand side of (1). The right hand side of (1) requires formulas for derivatives of $1/r$ in the direction n. These directional derivatives will be expressed in terms of derivatives in the x- and y-directions. The integration indicated by (1) is performed by summation according to Simpson's rule. The major elements into which the body is divided may vary from 10 to 150 or more. In the axially symmetric case it is necessary for purposes of integration to subdivide the body still further into subelements, the number depending principally upon the closest distance between an element j and the element i upon which p is located. When the points p and q coincide, the integrand in (1) becomes infinite. Integration for this singular subelement is accomplished analytically.

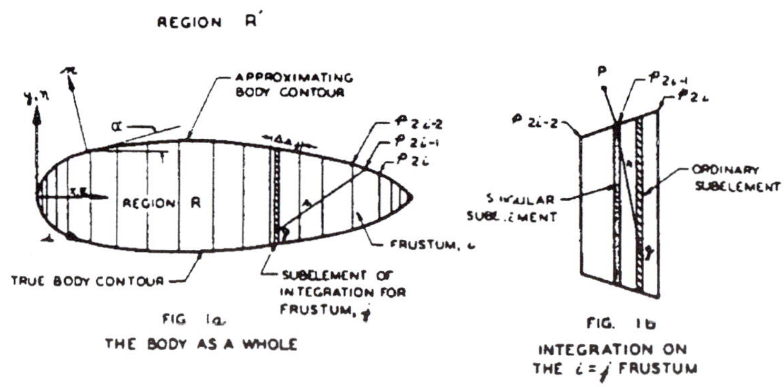

Figure1. Notation and method of approximating a body of revolution

The integral equation (1) is then solved for σ. If the body is divided into n elements, the integral equation can be converted into n simultaneous linear equations and solved algebraically for σ at the mid-point of each of the n elements. A modified form of Seidel's method has been used for solution of the system of equations. The order of the system of equations may be as large as 300, provided a large computer is used.

Major details of the method of solution follow, as applied to the more complicated case, that of axially symmetric flow. The two-dimensional problem is solved in substantially the same fashion but its solution is simpler because certain integrations not possible in the axially symmetric problem do become possible in the two-dimensional case. Because a homogeneous ring is the infinitesimal element of integration, the analysis begins by considering the potential and its derivatives for this shape. Further details may be found in [2].

2. Theory for a Homogeneous Ring

If λ is the linear density, and a is the radius of a homogeneous ring located as shown in Fig. 2, the potential at any point P in the x, y plane is

$$\varphi_P = 2\lambda \int_o^{\pi} \frac{a d\theta}{r} = 2a\lambda \int_o^{\pi} - \frac{d\theta}{\sqrt{x^2 + y^2 + a^2 - 2ay \cos \theta}} \tag{2}$$

Because it is permissible to differentiate under the integral sign, one can obtain expressions for the derivative of the potential in the two principal directions needed for computation of directional derivatives.

$$\left(\frac{\partial \varphi}{\partial x}\right)_P = - 2a\lambda \int_o^{\pi} \frac{a d\theta}{r} \int_o^{\pi} = - 2a\lambda \frac{x d\theta}{\left[x^2 + y^2 + a^2 - 2ay \cos \theta\ \right]^{3/2}} \tag{3a}$$

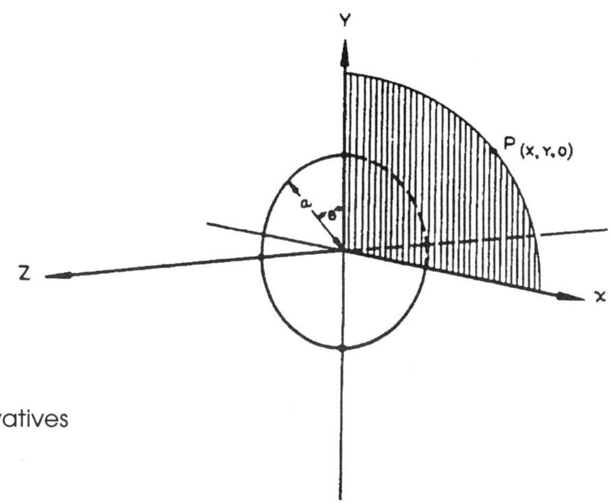

Figure 2.
Analysis of the potential and its derivatives due to a circular ring of radius a

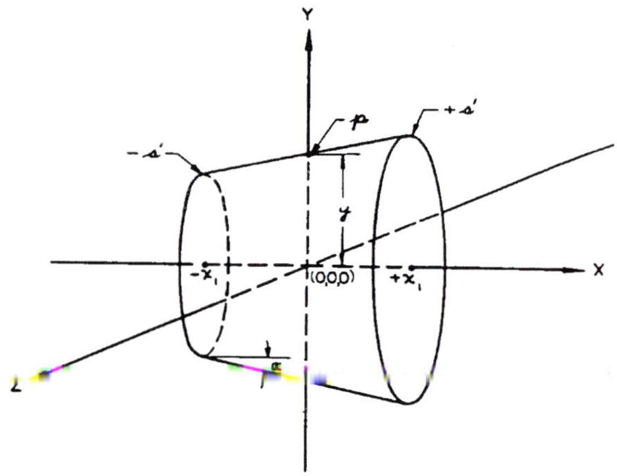

Figure 3. Diagram for analysis of the singular subelement of a frustum

$$\left(\frac{\partial \varphi}{\partial y}\right)_P = -2a\lambda \int_o^{\pi} \frac{(y - a\cos\theta)\, d\theta}{[x^2 + y^2 + a^2 - 2ay\cos\theta\ ]^{3/2}} \tag{3b}$$

Equations (3) can be converted to the forms below involving complete elliptic integrals. In these equations, since the infinitesimal element of integration over the surface of the body is a narrow ring of width δs whose source density is σ the quantity λ is replaced by its equivalent $\sigma\delta\sigma$. Moreover the ring element now is assumed to be located at $x = b$. Hence the contribution to the derivatives supplied by the infinitesimal ring element is

$$\delta\left(\frac{\partial \varphi}{\partial x}\right)_P = \frac{-4a\,(x+b)\,E\,(k)\,\sigma\delta s}{\sqrt{(y+a)^2 + (x-b)^2}\,[(y-a)^2 + (x-b)^2]} \tag{4a}$$

$$\delta\left(\frac{\partial \varphi}{\partial x}\right)_P = \frac{2a\sigma\delta s}{y\sqrt{(y+a)^2 + (x-b)^2}}\left[K(k) + \frac{y^2 - a^2 - (x-b)^2}{(y-a)^2 + (x-b)^2}\,E\,(k)\right] \tag{4b}$$

The quantity k is the modulus which here assumes the form

$$k^2 = \frac{4ay}{(y+a)^2 + (x-b)^2} \tag{5}$$

Equation (4) can be used to compute the contribution of any finite ring element at point P by simple numerical integration by dividing it into several subelements which, to be precise, are the lateral surfaces of frustums of cones.

2.1 The Derivatives of the Potential at the Point P when P and q Coincide—The Singular Subelement Case

Equation (4) is the basic integration formula, but it is not usable everywhere. When the point q coincides with P_{2i-1}, Fig. lb, the distance r becomes zero, k becomes equal to unity, K (1) is infinity, and so numerical integration near the singular point is impossible. The method of integration across the singularity is the subject of the present section.

Consider the frustum of a cone as illustrated in Fig. 3. Its slant height is equal to 2s′, and the point p is on the surface at the midpoint. Note that p is not a general point. The contribution to the derivatives of the potential lost as a consequence will be added later. Let the origin be so chosen that P is at (0, y, 0).

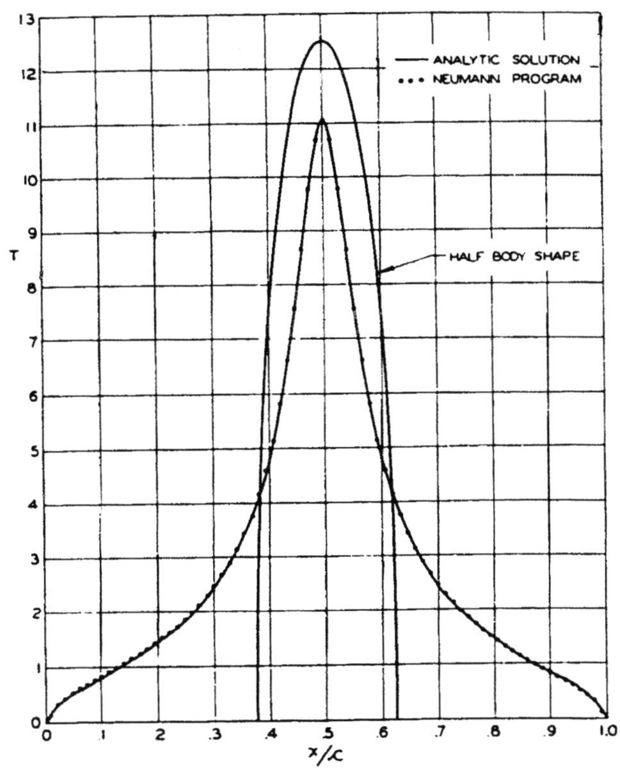

Figure 4. Velocity distribution on an elliptical cylinder having a fineness ratio of 0.1.two dimensional flow .76-point solution

Hence x equals zero, y is a constant, and a and b, (4), are variables equal to $y + s \sin \alpha$ and $s \cos \alpha$, respectively. Analytic integration becomes possible because the ring element may have a very small slant height 2s′.

We begin by specializing (4) and (5) to the present case and to the notation used in Fig. 3. We identify this partial contribution to the potential derivatives by the superscript 1. Then

$$k^2 = \frac{4(1 + (s/y) \sin \alpha)}{4 + 4(s/y) \sin \alpha + (s/y)^2} \tag{6}$$

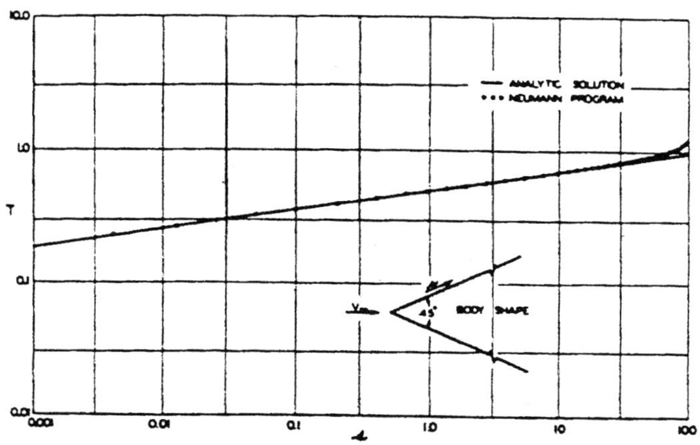

Figure 5. Velocity distribution along a 45° wedge. 61-point solution

$$\left(\frac{\partial \varphi}{\partial x}\right)^1_P = 4\sigma \cos \alpha \int_{-s'}^{s'} \frac{(1 + (s/y) \sin \alpha)\, E(k)}{\sqrt{4 + 4\,(s/y) \sin \alpha + (s/y)^2}} \; \frac{ds}{s} \tag{7a}$$

$$\left(\frac{\partial \varphi}{\partial x}\right)^1_P = -2\sigma \int_{-s'}^{s'} \frac{1 + (s/y) \sin \alpha}{\sqrt{4 + 4\,(s/y) \sin \alpha + (s/y)^2}} \left[K(k) - \frac{(s/y + 2 \sin \alpha)\, E(k)}{s/y}\right] \frac{ds}{y} \tag{7b}$$

In order to integrate (7), E and K must be expressed in terms of quantities that can be integrated analytically. Power series expansions can be used for the remaining troublesome terms, so that integrable expressions for both (7a) and (7b) can be obtained. Suitable expansions for the complete elliptic integrals, valid where the complementary modulus k′ is small are, with $\Lambda = \ln (4/k')$

$$E = 1 + \frac{1}{2}\left(\Lambda - \frac{1}{2}\right)k'^2 + \frac{3}{16}\left(\Lambda - \frac{13}{12}\right)k'^4 + \frac{15}{128}\left(\Lambda - \frac{6}{5}\ k'^6\right) + \ldots \quad (8)$$

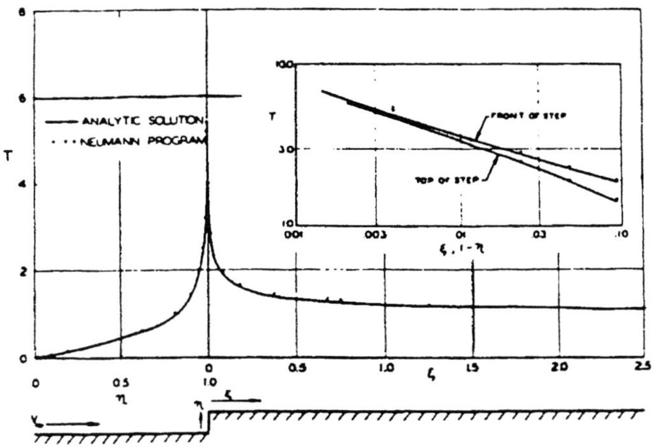

Figure 6. Flow past a square-cornered step. 41-point solution. The auxiliary plot shows the velocities very near the corner.

$$K = \Lambda + \frac{\Lambda - 1}{2}k'^2 + \frac{9}{64}\left(\Lambda - \frac{7}{6}\right)k'^4 + \frac{25}{256}\left(\Lambda - \frac{37}{30}\right)k'^6 + \ldots \quad (9)$$

By means of these two equations plus (6) and the procedure just outlined, (7) can be transformed to a series of terms of the type S^n or $S^n \ln S$, where $S = s/y$. Clearly integration now is elementary, and the result is with $S' \equiv s'/y$:

$$\left(\frac{\partial \varphi}{\partial x}\right)_\gamma^1 = \sigma \sin\alpha \cos\alpha \left[2S' + \left(\frac{13}{72} + \frac{1}{24}\ ln\left(\frac{S'}{8}\right) + \frac{1}{12}\sin^2\alpha\right) S'^3 + \ldots\right] \quad (10a)$$

$$\left(\frac{\partial \varphi}{\partial x}\right)_p^1 = \sigma \left[\left(2\sin^2\alpha + 2\ ln\left(\frac{S'}{8}\right)\right)S' - \frac{1}{24}\left(3 + 3\ ln\left(\frac{S}{8}\right) - 3\sin^2\alpha - sin^4\alpha\right) S' + \ldots\right] \quad (10b)$$

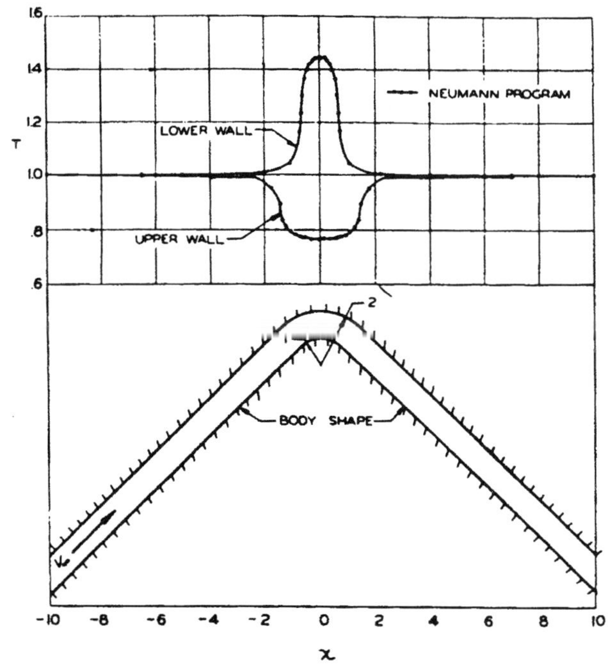

Figure 7. Velocity distribution along walls of a two-dimensional 90° bend. 65-point solution

Equations (10) are the final results for the singular subelement, except that the quantity $2\pi\sigma$ is missing. Potential theory has thoroughly established the fact that the normal derivatives of the potential at a point p for exterior and interior regions are respectively.

$$\frac{\partial \varphi}{\partial n_+}(p) = -\,2\pi\sigma(p) + \iint_S \sigma(q)\frac{\partial}{\partial n}\left(\frac{1}{r}\right) dS$$

$$\frac{\partial \varphi}{\partial n_-}(p) = +\,2\pi\sigma(p) + \iint_S \sigma(q)\frac{\partial}{\partial n}\left(\frac{1}{r}\right) dS$$

Because of the way (10) was derived, the quantity $2\pi\sigma(p)$ is missing, and must therefore be added in order to obtain the entire contribution of the surface density distribution. In the original work, expressions for the more general point off the surface were derived and the additional quantity $2\pi\sigma$ was indeed obtained.

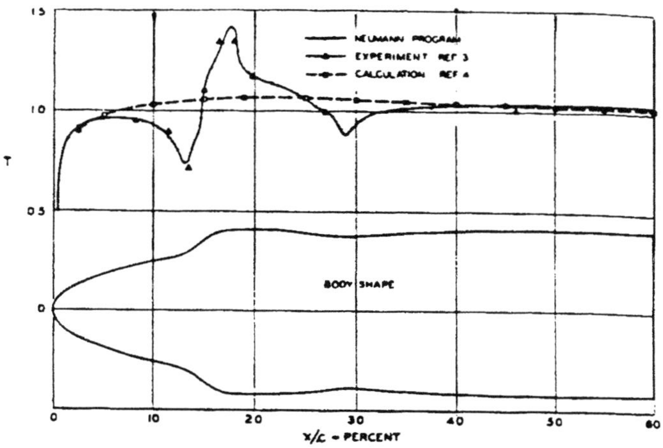

Figure 8. Experimental and theoretical velocity distribution at zero angle of attack over a prolate spheroid having an annular bump.

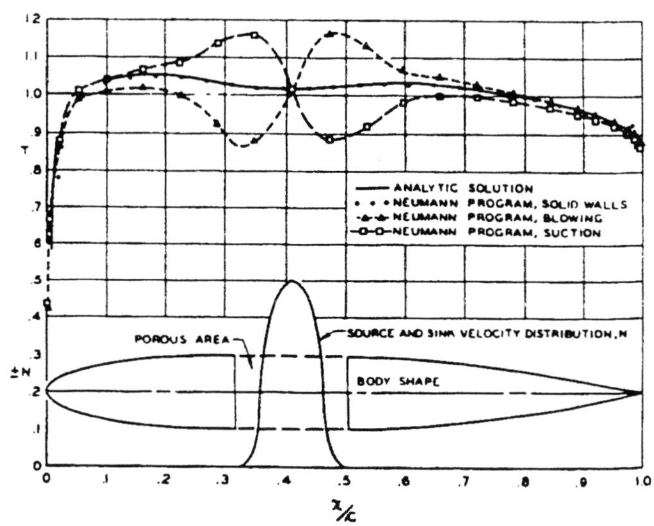

Figure 9. Effect of area suction or blowing near center of a body of revolution. 25-point solution.

But in the interests of clarity, the more complicated derivation has not been presented here.

Then when the normal derivative is resolved into its components we have for the singular subelement when dealing with an external flow.

$$\left.\begin{aligned}\frac{\partial \varphi}{\partial n_+} &= 2\pi\sigma_+ \sin\ \alpha + \left(\frac{\partial \varphi}{\partial n}\right)^1 \\ \frac{\partial \varphi}{\partial y_+} &= -\ 2\pi\sigma_+ \cos\ \alpha + \left(\frac{\partial \varphi}{\partial y}\right)^1\end{aligned}\right\}\ \text{Singular Subelement}$$

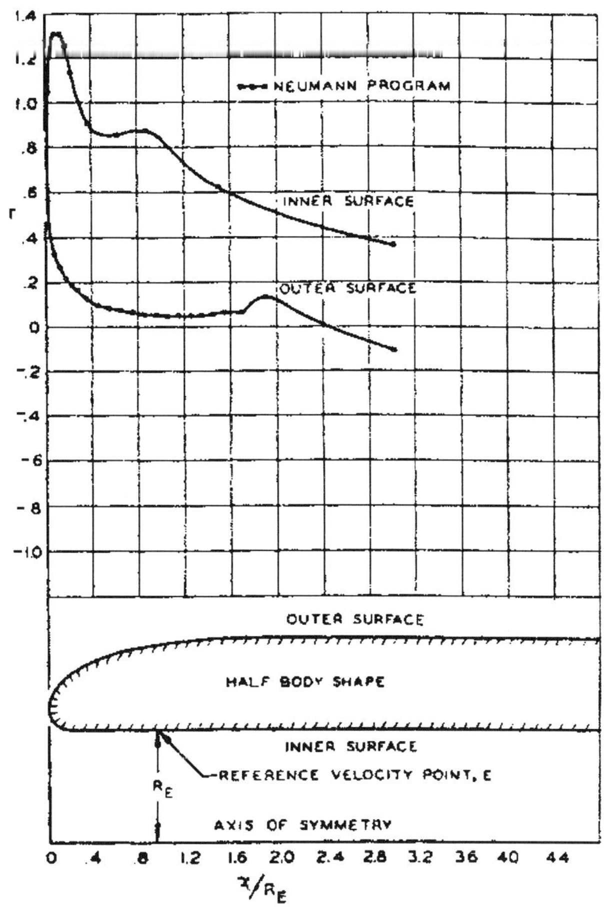

Figure 10A Velocity distribution about an axially symmetric inlet. 41-point solution. First independent solution.

4. Formulation of the System of Equations

The necessary formulas are now available for determining the derivative of the potential in any direction at any point on the body. Equations (4) cover most cases and (10) cover the remaining, the cases when a subelement becomes singular. Hence we are now in a position to formulate a system of equations for the unknown values of σ .

Because the method of solution of the integral equation (1) is algebraic, it now becomes necessary to identify the various elements. The general system of identification is indicated in Fig. 1. The letter *i* corresponds to *p*, the point of interest, and *j* corresponds to *q*, the variable of integration. Let s_{2i-2}, s_{2i} be the *s*-coordinates of the end points and s_{2i-1} the midpoint of the *i*-th frustum. In Fig. 1 these points are labeled p_{2i-2}, p_{2i}, and p_{2i-1}, respectively. The Cartesian coordinates of these points are (x_{2i-2}, h_{2i-2}), (x_{2i}, h_{2i}), and (x_{2i-1}, h_{2i-1}) respectively. In the section dealing with the frustum $i = j$, the quantities x , h were represented by the symbols *b, a*. Neglecting for the moment the quantity $2\pi\sigma$, the surface density of an element *j* provides values for the derivatives of the potential at any midpoint p_{2i-1} which can be written thus as perhaps a clearer restatement of (4) or (10):

$$\left(\frac{\partial\varphi}{\partial x}\right)^1_{ij} = \sigma_j X_{ij} \qquad \left(\frac{\partial\varphi}{\partial y}\right)^1_{ij} = \sigma_j Y_{ij}$$

The complicated quantities X_{ij}, Y_{ij} are determined solely by the geometry of the body. Then according to (4) which are applicable everywhere but on the singular subelement, and with assignment of identifying subscripts, we have for each individual frustum *i*.

$$\underset{(i\neq j)}{X_{ij}} = -4\int_{s_{2j-2}}^{s_{2j}} \frac{\eta_j(s_{2i-1}-\xi_j)\,E(k)\,ds}{\sqrt{(y_{2i-1}+\eta_j)^2+(x_{2i-1}-\xi_j)^2}\,[(y_{2i-1}-\eta_j)^2+(x_{2i-1}-\xi_j)^2} \tag{11a}$$

$$\underset{(i\neq j)}{Y_{ij}} = -2\int_{s_{2j-2}}^{s_{2j}} \frac{\eta_j}{y_{2i-1}\sqrt{(y_{2i-1}+\eta_j)^2+(x_{2i-1}-\xi_j)^2}}\cdot \tag{11b}$$

$$\left[K(k) + \frac{y_{2i-1}^2-\eta_j^2-(x_{2i-1}-\xi_j)^2}{(y_{2i-1}-\eta_j-1)^2+(x_{2i-1}-\xi_j)^2}\,E(k)\right]ds$$

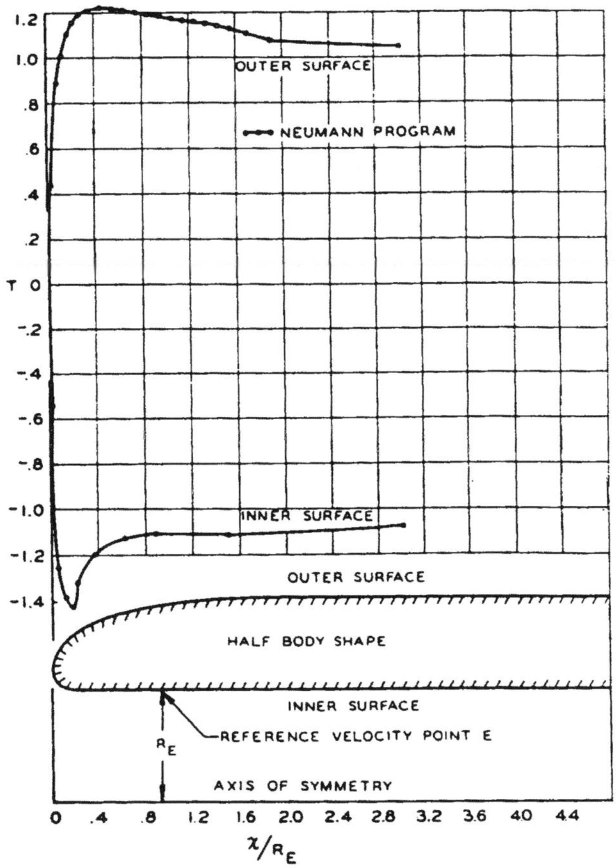

Figure 10B. Velocity distribution about an axially symmetric inlet. 41-point solution. Second independent solution.

The integration over the complete boundary is performed numerically by Simpson's rule. Machine subroutines are available for evaluation of E and K. When $i = j$, both (4) and (10) are used to account for the singular element, and the following formulas are obtained where $S'_i = s'/\gamma_{21-1} = \Delta s/2\gamma_{2i-1}$.

(12b)

$$X_{ij} = \sin \alpha_i \cos \alpha_i \left[2S_i' + \left(\frac{13}{72} + \frac{1}{12} \ln (S_i'/8) + \frac{1}{12} \sin^2 \alpha_i \right) S_i'^3 + \ldots \right] \tag{12a}$$

$$- 4 \int_{s_{2i-2}}^{s_{2i-1} - s_i' y_{2i-1}} \frac{\eta_i (x_{2i-1} - \xi_i) \, E(k) \, ds}{\sqrt{(y_{2i-1} + \eta_i)^2 + (x_{2i-1} - \xi_i)^2} \left[(y_{2i-1} - \eta_i)^2 + (x_{2i-1} - \xi_i)^2 \right]}$$

$$- 4 \int_{s_{2i-1} + s_i' y_{2i-1}}^{s_{2i}} \frac{\eta_i (x_{2i-1} - \xi_i) \, E(k) \, ds}{\sqrt{(y_{2i-1} + \eta_i)^2 + (x_{2i-1} - \xi_i)^2} \left[(y_{2i-1} - \eta_i)^2 + (x_{2i-1} - \xi_i)^2 \right]}$$

$$Y_{ii} = (2 \sin^2 \alpha_i + 2 \ln (S_i'/8)) S_i' - \frac{1}{24} (3 + 3 \ln (S_i'/8) - 3 \sin^2 \alpha_i - 2 \sin^4 \alpha_i) S_i'^3 + \ldots \tag{12b}$$

$$- 2 \int_{s_{2i-2}}^{s_{2i-1} - s_i' y_{2i-1}} \frac{\eta_i}{y_{2i-1} \sqrt{(y_{2i-1} + \eta_i)^2 + (x_{2i-1} - \xi_i)^2}} \cdot$$

$$\left[K(k) + \frac{y_{2i-1}^2 - \eta_j^2 - (x_{2i-1} - \xi_i)^2}{(y_{2i-1} - \eta_i)^2 + (x_{2i-1} - \xi_i)^2} E(k) \right] ds$$

$$- 2 \int_{s_{2i-1} + s_i' y_{2i-1}}^{s_{2i}} \frac{\eta_i}{y_{2i-1} \sqrt{(y_{2i-1} + \eta_i)^2 + (x_{2i-1} - \xi_i)^2}} \cdot$$

$$\left[K(k) + \frac{y_{2i-1}^2 - \eta_i^2 - (x_{2i-1} - \xi_i)^2}{(y_{2i-1} - \eta_i)^2 + (x_{2i-1} - \xi_i)^2} E(k) \right] ds$$

Finally, since $\partial\varphi/\partial n$ along the boundary is known, we see from (1) that the system of simultaneous equations becomes for external flows

$$\left(\frac{\partial\varphi}{\partial x}\right)_{2i-1} = -2\pi\sigma_i - \sin\alpha_i \sum_{j=1}^{n} \sigma_j X_{ij} + \cos\alpha_i \sum_{j=1}^{n} \sigma_j Y_{ij} \qquad \text{(13a)}$$

Once the system (13a) is solved for the n-values of σ_i, velocities tangential to the body contributed by the surface density distribution can be computed by the formula

$$\left(\frac{\partial\varphi}{\partial s}\right)_{2i-1} = \cos\alpha_i \sum_{j=1}^{n} \sigma_j X_{ij} + \sin\alpha_i \sum_{j=1}^{n} \sigma_j Y_{ij} \qquad \text{(13b)}$$

It is shown in [2] that (13) is applicable equally to both internal and external flows, provided the profile defining the body is traversed in such a way that the flow field of interest is kept to the left.

If the body has an impermeable wall, and is moving through a stationary fluid with a velocity V_∞, $\partial\varphi/\partial n$ at the wall is $V_\infty \sin\alpha_i$. Then the final form of the system (13a) defining the distribution of σ is

$$V_\infty \sin\alpha_i = -2\pi\sigma_i - \sin\alpha_i \sum_{j=1}^{n} \sigma_j X_{ij} + \cos\alpha_i \sum_{j=1}^{n} \sigma_j Y_{ij} \qquad \text{(14)}$$

The final complete tangential velocity is computed by adding the free stream component to the value supplied by (13b). Velocities at points in the stream can be computed by application of (4) alone, since the singularity disappears once the surface of the body is left.

The two-dimensional solution is derived generally in the same fashion, but the analysis is easier. For it the elements become planes of constant density instead of frustums of cones. Analytic integration across the element becomes possible and therefore it is unnecessary to break up the element into subelements. Furthermore, since the integration is possible for the general point P, the quantity $2\pi\sigma$ is automatically included in the final formulas.

5. Examples and Discussion for Two-Dimensional Flows

The method of solution just presented is quite powerful. For example, semi-infinite bodies can be treated merely by terminating the table of coordinates at a

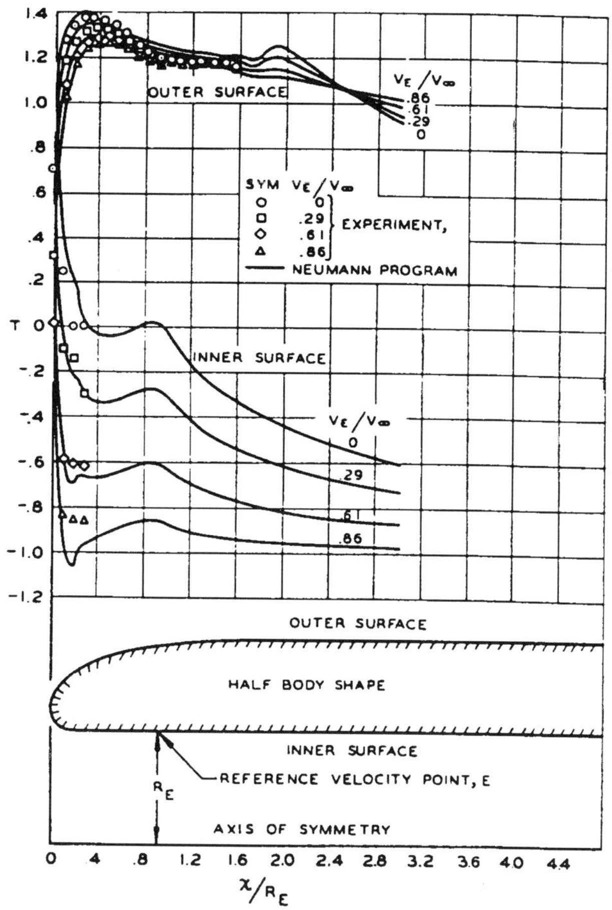

Figure 10C. Velocity distribution about an axially symmetric inlet. 41-point solution. Combined solutions.

reasonable distance away from the area of interest. Flow around sharp corners is found to be calculable and to approach infinity correctly. Flow through permeable walls presents no complication; all that is necessary is to modify $\partial\varphi/\partial n$ to account for the desired amount of flow through the walls. Because there are no restrictions in the analysis that limit computations to a single body, flows about aggregates of several bodies can be calculated. The main limitations at the present time are that (a) there can be no lift, (b) the flows must be two-dimensional or axially symmetric, and (c) the body or bodies must be

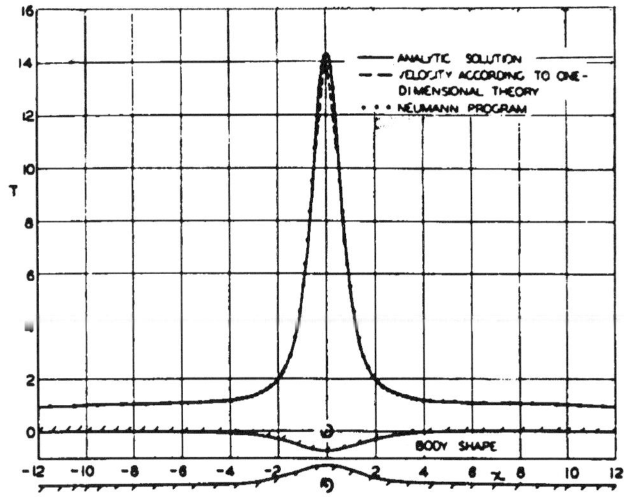

Figure 11. Flow through a ring vortex construction. 81-point solution.

definable by not more than about 300 coordinate points, assuming computation is done on the IBM 704 or equivalent. Flows about several hundred different shapes have been calculated; results for a few are shown on the following pages. The shapes exhibited are selected principally to indicate both the accuracy of the method and the diversity of shapes that have been treated.

Elliptic cylinder having a thickness ratio of 10:1

Fig. 4 shows a 76-point solution of the flow past a very thick ellipse. An ellipse was chosen for the reason that it has exact solutions. In this and all subsequent figures all data points are plotted in order to indicate both their number and their distribution. The present method is identified as the "Neumann program." Errors are small and the peak velocity is predicted with high accuracy. The tangential velocity ratio is denoted by the symbol *T*.

Velocity distribution over a semi- infinite 45⁰ wedge

Early it was learned that ordinary semi-infinite bodies similar to source bodies presented no problem. In their case the velocity ratio *T* approaches unity far downstream from the nose. Curiosity developed as to whether two-dimen-sional-wedge flows could be calculated. For them the velocity on the surface is $T = ks^{b/(2-b)}$ where bp is the entire wedge angle. That is, if $\beta > 0$, the velocity T now goes to infinity downstream, as does the body width itself. Results of a 61-point calculation for a 45^0 wedge are shown in Fig. 5. High accuracy is maintained over three full log cycles.

Flow past a square-cornered step

In many problems discontinuities in slope exist. An outstanding example is the trailing edge of an airfoil where infinite velocities would usually occur were it not for circulation. Another question is whether or not a single boundary can be treated, that is, whether or not flows in the upper half plane along an arbitrary infinite boundary can be computed. Figure 6 represents a severe test of both points. It shows the solution for an infinitely deep body of fluid streaming past a square-cornered step. The wall from minus infinity to plus infinity is taken to be the boundary. The problem could also be solved by treating the step as the front of a square-nosed semi-infinite body. The exact solution is supplied by a Schwartz-Christoffel transformation.

The results can be seen in Fig. 6; although only 41 points were used, the accuracy is high. Of particular interest is the manner in which velocities near the corner approach infinity. The insert in the figure is an enlarged logarithmic plot of velocities very close to the corner. In view of the fact that no great effort was made to get accurate results near the corner, the results are good. The solution required about 12 minutes on the IBM 701 Calculator.

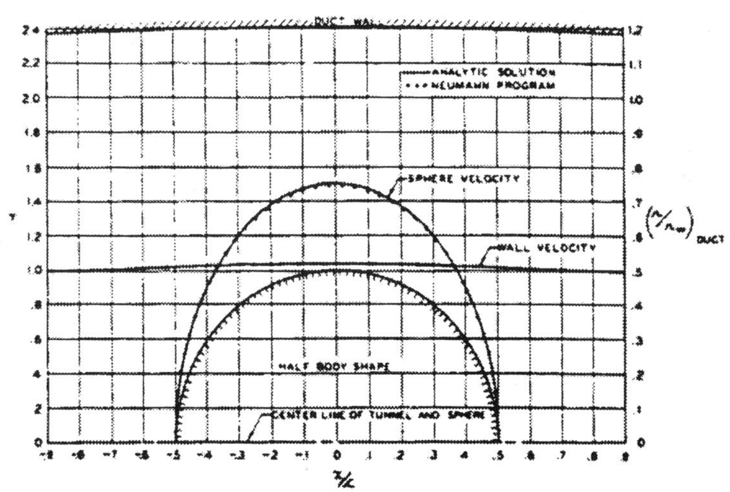

Figure 12. Flow past a sphere in a circular variable-area duct. 81-point solution.

Velocity distribution along walls of a 90⁰ elbow

Figure 7 shows the velocity distribution along the walls of a 90^0 elbow. The boundary conditions were established by treating the bend as if it were moving normal to its plane of symmetry through still air. The velocity distribution has very strong adverse gradients that undoubtedly cause separation and consequent high bend loss. No exact solutions are available.

6. Examples of Axially Symmetric Flows

Flow past a bumpy body

Fig. 8 is an example of an axially symmetric shape previously insolvable. The fore part of the body is half of a prolate spheroid upon which is superimposed the axially symmetric bump as shown. Pressure distribution measurements made on a wind tunnel model are shown as the solid triangles. For comparison pressure distribution calculated by one of the best current methods is shown by the open squares. The method breaks down completely; it fails entirely to respond to the bump. The agreement between experiment and the present theory is as good as could be expected for such a bumpy shape when boundary-layer effects are considered.

Velocity distribution on a body of revolution having area suction

To learn the effect of distributed inflow and to search for calculation difficulties, a test case was devised. The body used was a special source-sink body whose contour and velocity distribution were known to 4-place accuracy. Near the center a strong sinusoidally varying normal flow N was introduced, as shown in Fig. 9. Three cases are shown on this figure: the solid-wall solution, suction, and blowing solutions. The solid-wall solution is in excellent agreement with the analytic solution and can be regarded as a test case for a conventional streamline shape. Because of the complete unavailability of theoretical calculations and because the meager test results available are inadequate, no statement as to accuracy can be made for the two suction cases.

Flow past on inlet

Inlets represent the only shape giving difficulty in calculation. Apparently when the walls are thin, the inner and outer surfaces are so close to each other that the system of simultaneous equations is poorly conditioned and Seidel's method of solution becomes slow or may even fail. No trouble is met dealing with thicker inlets of the type illustrated in Fig. 10, other than the fact that these more complicated shapes require more points for definition. The entire range of solutions is obtained by linear combinations of two independent solutions. One of these, corresponding to the inlet at rest and having a mass flow ratio of infinity, is shown in Fig. 10a. The necessary inflow rate is obtained merely by the artifice of specifying the velocity across a boundary located some distance down the duct. This device—that is, a sink sheet—for inducing flow has earned, appropriately enough, the name of 'pump.' Fig. 10b shows the second independent solution. It represents the inlet with a clear duct, advancing through the air in normal fashion, without any control on the amount of internal flow. The combined solutions are shown in Fig. 10c. The disagreement with experimental data is no greater than the experimental error. In Fig. 10a the "bump" in the velocity distribution some distance aft of the nose should be ignored, for it is caused by the "pump."

A ring vortex constriction

Exact solutions of internal flow problems are most rare. Therefore as a test problem a ring vortex was superimposed upon a uniform stream to obtain a constriction in a round tube. The shape and velocity distribution were calculated to 4- place accuracy; these are shown in Fig. 11. As can be seen from the figure, the 81-point solution has very high accuracy. The velocity ratio computed by one-dimensional theory is added for its interest.

A sphere in a circular, variable-area duct

Again a synthetic problem must be created if we demand the exact solution of a test case dealing with wall effects on a body of revolution located on the axis of a channel of circular cross section. It was decided to use a sphere in free air as a test case. The position of one of the streamlines was calculated. This streamline was in turn treated as the tunnel wall, and the flow about the sphere was calculated. Because the wall coincides with one of the streamlines, the free-air velocity distribution both along the sphere and along the tunnel wall should be recovered. The Neumann and the exact solutions are compared in Fig. 12. Again the accuracy is good. This example is rather complicated, but machine capacity was not taxed, because good accuracy was obtained with only 81-points.

Bibliography

1. Kellogg, 0. D.: Foundations of Potential Theory. Frederick Ungar Publishing Co., 1929, also available through Dover Publications, Inc.

2. Smith, A. M. 0. and Pierce, Jesse.: Exact Solution of the Neumann Problem. Calculation of Non-Circulatory Plane and Axially Symmetric Flows About or Within Arbitrary Boundaries. Douglas Aircraft Co. Rept. No. ES26988, 1958.

3. Matthews, Clarence W.: A Comparison of the Experimental Subsonic Pressure Distribution About Several Bodies of Revolution with Pressure Distributions Computed by Means of Linearized Theory. NACA RM L9F28, Sept.12, 1944.

4. Young, A. D., and Owen, P. R.: A Simplified Theory for Streamline Bodies of Revolution, and its Application to the Development of High-speed Low-drag Shapes. British Ministry of Supply R and M No.2071, July 1943.

AMO and Betty in 1974

5

Boundary layers

by
Tuncer Cebeci
A member of AMO's Research Group

From my personal viewpoint, my education truly began with my association with AMO. As he was always fond of observing "Tunc didn't know beans about boundary layers when he joined Douglas. (In my defense, that was not completely true because I had a few beans in my bag!) I also knew very little about airplanes and reluctantly experienced my first flight in order to reach my golden opportunity in 1965. I had moved my family from North Carolina and arrived (with great excitement and pride that AMO Smith had hired ME) only to find that he not only did not recognize me, but did not recall hiring me! Luckily, he told me I could stick around but the next hurdle presented itself: he could not remember my name. I changed the Turkish pronunciation of my surname to a phonetic one — CBC. Then, since it was lunch time, I created my first name - "Tunc like 'lunch.' " Thus on my first day with AMO, he created my professional name.

AMO worked in many aspects of boundary layer flows. He first started the calculation of laminar boundary layers by improving the solutions of the Falkner and Skan equation in 1954 (ref. 4); he then developed a rapid laminar boundary layer method base don the piecewise application of similar solutions (ref. 11). His work on the numerical solutions of the nonsimilar laminar boundary layer equations began with Darwin Clutter around 1962 and was followed by a publication in the AIAA J., September 1963 (ref. 19). In

December 1963 he was asked by Science Citation index to write a short article about this work because it had received many citations (ref. 19).

His work on the calculation and modelling of turbulent boundary layers began around 1964 and was followed by a paper in June 1966 (ref. 27). Under his guidance, I started working with him around that time, refining the calculation method and improving the modelling of turbulent flows. With Herb Keller, a tamed and very gifted mathematician from Cal Tech hired by AMO as a consultant, we developed modern numerical methods for solving the laminar and turbulent boundary layer equations for incompressible and compressible flows (refs. 35, 37, 42-45, 47, 48, 51, 52, 54, 55). In 1974, AMO and I published a book entitled "The Analysis of Turbulent Boundary Layers," (ref. 57). Based on the procedures AMO and I developed, his approach to boundary layer flows has led to the development of an interactive boundary layer method, which is another approach to solving the Navier-Stokes equations for many aerodynamic flows. In all of my work in boundary layer flows, his guidance and his way of approaching the problems played a major inspiring role.

THE BOUNDARY LAYER AND I*

by

A. M. O. Smith

Nomenclature

A = amplitude of Tollmien-Schlichting waves or Görtler vortices; also Van Driest's damping- length constant
A_0 = amplitude at neutral stability
c = constant in velocity distributions of the form $u_e = cx^m$
C_{Dp} = profile drag coefficient
F''_W = transformed value of $\partial u/\partial y$ at the wall in the Falkner-Skan equation
k = height of a roughness
m = exponent in velocities of the form $u_e = cx^m$
n = measure of amplification ratio, $n = \ln (A/A_0)$
Pr = Prandtl number
r = longitudinal radius of curvature for Görtler vortices
R_c = chord Reynolds number
R_θ = Reynolds number based on momentum thickness
u = velocity in x direction
u_e = edge velocity
x,y = coordinates, generally along and perpendicular to the wall
β = Hartree's boundary-layer profile parameter, note $\beta = 2m/(m+1)$
γ = intermittency factor
δ = thickness of a boundary layer
$\delta*$ = displacement thickness
ε = turbulent eddy viscosity; ε_i, inner, ε_o, outer
θ = momentum thickness of a boundary layer
k = von Kármán's constant
ν = kinematic viscosity
ρ = mass density
Λ = Pohlhausen's $\Lambda = (\delta^2/\nu)(du_e/dx)$

1. Early Motivation

My story begins in June 1945, in a basement at the Göttingen Aerodynamische Versuchsanstalt (AVA), Germany, at the end of World War II. But first let us back up slightly. In 1937 and 1938, I took a course, called AE

*AIAA J., Vol. 19, No. 11, November 1981, p. 1377

267 - Real Fluids, at Caltech from C. B. Millikan. This course dealt with laminar and ordinary viscous flows. AE 269 dealt with turbulent flows. AE 267 was rather similar to the laminar and viscous portions of Goldstein's *Modern Developments in Fluid Dynamics* and Prandtl's contribution in Durand's *Aerodynamic Theory*, Vol. III. It was a very good course and Millikan was probably the clearest and best organized teacher I ever had. I know that he always conscientiously reviewed his next day's lectures the day before. However, I was mainly interested in becoming a preliminary designer, and at that time, boundary layers and viscous effects seemed quite longhaired and abstruse.

Part of the course dealt with the Orr-Sommerfeld equation governing the stability of laminar boundary layers and attempts at the solution of the equation. This was long before Schubauer and Skramstad verified its applicability to the process of boundary-layer transition, so the analysis was a kind of academic exercise. At one time in some general discussion, Millikan mentioned that suction should change the shape of the boundary-layer profile and probably make it more stable, but nothing was known about how much suction it would take or whether there could be net gains in efficiency, such as net gains in total required horsepower. After leaving Caltech, I did not remember much about this theory but did remember his conjecture regarding stabilization by suction and the question about how much suction might be required.

My work at Douglas Aircraft from 1938 to 1945 was on general airplane design and ordinary aerodynamics, and I never thought about the subject of laminar-flow control, although my Master's thesis had been on high-lift boundary-layer control. However, in 1945 at the end of the war, I was appointed a member of the U.S. Naval Technical Mission in Europe (NAVTECMISEU) to help investigate the German aeronautical developments. One part of my work involved exploring the Göttingen AVA and Kaiser Wilhelm Institute (KWI) complexes. I got into their reprint room in a basement and browsed through many interesting new and old and famous classical papers. One that I came across was by Schlichting and Bussmann.[1] This paper solved the Falkner-Skan equation for suction and blowing type of boundary conditions on the wall. The authors found that the stability would be changed tremendously by very small amounts of suction. Thus, it answered the question posed by Millikan seven years before and caused me to pay serious attention to laminar-flow control as a possibility for drag reduction. However, I did not immediately try to do any work in this area and nothing happened, except that I followed Pfenninger's Swiss work. Then it was my good luck to be in Washington in December 1947 at the time Sidney Goldstein gave his

Wright Brothers lecture about the Griffith airfoils developed in England. A typical Griffith airfoil is illustrated in Fig. 1. It has long, continuously favor-

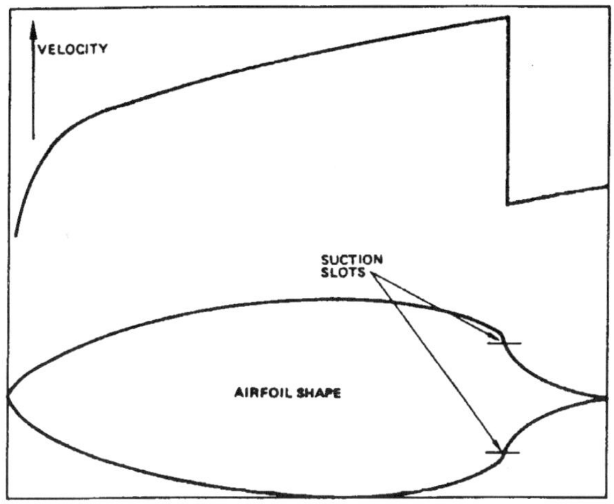

Figure 1. A typical thick symmetric Griffith airfoil, $\alpha = 0$ deg.

able pressure gradients on each side. In order to close the airfoil, there must be a pressure jump toward the rear, and after that, the flow again accelerates. At the pressure jump, suction is applied to prevent separation. One of the unique features is that, behind the slots, the shape must be concave as seen in the illustration in order to develop the higher pressures. While listening to the lecture, it occurred to me that an airfoil could be designed with several small pressure jumps instead of one large one on each side—perhaps 10 per side. My design would lead to an airfoil that had a more conventional shape, and in case suction failed, its performance should not deteriorate catastrophically. I wanted to find something better than Pfenninger's approach of using a large number of slots (40 per side or more) and its resultant mechanical complications.

But as already noted, when I learned of the results of Ref. 1, I did not immediately try to start work on laminar flow airfoils. In 1948, I believe, I attended a Rand Corporation briefing on ways to extend the range of aircraft. One way was by aerial refueling. This seemed very wild to me and if people were that desperate to extend the range, then laminar flow control should surely be looked into as an alternate. After that briefing I began active work in this area, mainly around my idea. Then in attempting to design practical and successful Griffith type airfoils I was forced into experimental and theoretical boundary-layer work rather extensively.

2. DESA-2 Airfoil and Theoretical Developments

This airfoil was the type I conceived at Goldstein's lecture. It was not successful, but it was a very basic stimulus to much of my work in the field, so I feel it deserves separate mention. We first designed and tested a preliminary model, the [illegible] airfoil. It was symmetrical and had a flap. We obtained 98% laminar flow to the capacity of the wind tunnel, $R_c = 4.25 \times 10^6$. At that time this was the highest Reynolds number for full laminar flow that yet had been reached. The tests were a complete success. Therefore, we began designing a thin, cambered and more realistic shape for tests up to $R_c = 40$ or 50×10^6 in the NACA two-dimensional pressure tunnel. It was 6.6% thick. The shape and pressure distribution are shown in Fig. 2. Just behind each

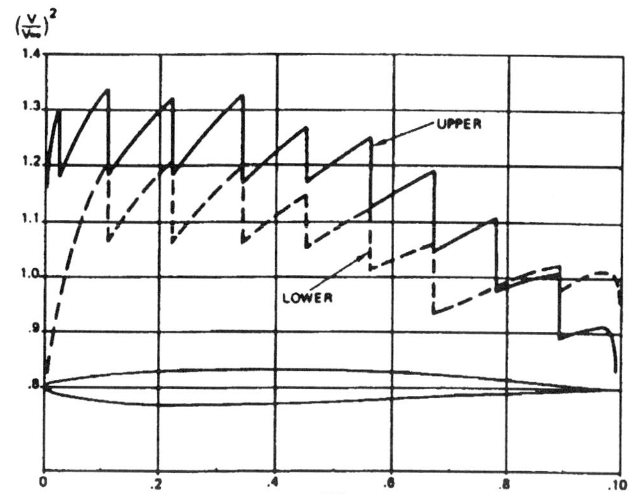

Figure 2. The DESA-2 airfoil and its pressure distribution, 6.6% thick, design $C_L = 0.1$.

pressure jump is a short concave region. At the scale of the figure, the waviness and concavities required to generate the sawtooth pressure distribution do not show. Moreover, the pressure jumps are so low that, in case of suction failure, the shape should act like an ordinary airfoil. In wind tunnel tests, both the $G0010_7$ and the DESA-2 did so behave.

The model has two unique features: 1) very strong stabilization due to the very favorable pressure gradients and 2) several short regions of substantial concave curvature behind the slots. The concave curvature was substantial, but not so great as to invalidate boundary-layer theory. By this, I mean that, in the basic equations, δ/r is still small. Here, δ is a measure of the boundary-layer thickness, and r the longitudinal concave radius of curvature. Moreover, the regions were very short. In thinking about this situation as well as the favorable gradient regions, it did not seem reasonable to me that transition would be determined by an instantaneous value of some parameter. To change the flow, some process had to act for a finite period, because physical processes are never truly impulsive, and there is nothing to give an impulse. It seemed to me that any analysis of the process should involve an integration, not just a limiting value, unlike an allowable stress in the science of strength of materials.

The concave regions would develop Görtler vortices, which if too strong, would cause transition. Görtler's work dealt only with neutral stability; however, in our problem, there would be strong instability, but only for a very short time. Hence, we wanted to know rates of growth, and Görtler's paper gave no indication of rates. Thus, this peculiarity of the DESA-2 directly involved me deeply in boundary-layer theory and made me feel that an amplification ratio was more significant than the maximum value of some parameter, e.g., Görtler's $R_\theta(\theta/r)^{1/2}$. Here θ is momentum thickness, and R_θ is the usual momentum thickness Reynolds number. Furthermore, it should be noted that the stability equations yield rate parameters, not strength parameters. Strength of a disturbance is some kind of integral of rate times distance, or rate times time.

During the design of the DESA-2 airfoil, I often stated that I was not sure the airfoil would work, but at least I was determined not to have to apologize for a poor job of designing it. Therefore, I decided to try to calculate the growth rates of Görtler vortices. For me this was a substantial effort, because, up to that time at Douglas, I had hardly used any mathematics beyond ordinary algebra. I studied various methods of solving the eigenvalue problem, including the basic mathematics. I remember studying Pauling and Wilson's *Introduction to Quantum Mechanks* and Margenau and Murphy's book, *The Mathematics of Physics and Chemistry*. Also, you might say Ince's *Ordinary Differential Equations* was my bible. In the boundary-layer field, I studied both Meksyn's work, G. I. Taylor's, and Görtler's own. In fact, we solved several cases by

Görtler's own method for neutral stability, but it did not seem to have the power needed for solving the problem we wished to solve, which involved amplification rates rather than just neutral stability. As will be seen, we did find what we thought was a more powerful method.

However, one problem bothered me. I could not tie time and distance together, as can be done with Tollmien-Schlichting waves, and of course, Görtler's and Meksyn's analyses were for wave growth in time. I pondered this problem for some months and one weekend, while laying flagstones in my backyard, saw the solution. Instead of using a disturbance expression that grew as $\exp t$, the problem should be reformulated into one that grew as $\exp x$.

At about the same time, late 1949 1 believe, I talked with Douglas Dill, head of our Douglas El Segundo Flutter Group. I was looking for help on methods of calculation. He suggested that I look into Galerkin's method. I did and found it so powerful that it could nicely handle the kind of problem we wished to solve. This method involves solution of a matrix. Mechanical automatic calculators, which were good at matrices, were just arriving, and this was more incentive for using Galerkin's method.

Accordingly, we worked hard on deriving the extended equations and studying their structure, order of magnitude of various terms, methods for calculation of eigenvalues, etc. Frazier, Duncan, and Collar's book, *Elementary Matrices*, was my source book for matrix methods and theory. I was forced to gain a great deal of knowledge or I would never have solved our problem. The unusual nature of the DESA-2 and the high-performance stakes were the stimulus. At the design $R_c = 50 \times 10^6$, the predicted effective drag coefficient was C_{D_p} = 0.00064. For comparison, wind tunnel tests show the drag of the NACA 66, 1-015 airfoil to be C_{D_p} = 0.0050 at this same Reynolds number. Therefore, as you see, the DESA-2 airfoil and its accompanying stability problem caused me to pay much more attention to boundary layer and applied mathematics literature.

As an aside, it should be noted that the calculations we ended up doing can only be called "heroic." While we studied accuracy with lesser matrices, for our final results, we set up a 12 x 12 matrix system, expecting the IBM machine to give us the answers. When we gave the operator the first case, we started solving it on a desk calculator as a check. The process required inversion of the matrix and repeated multiplication by a column to get the eigenvalue. Also, the matrix was nearly singular, i.e., very difficult. The man who attacked the problem with a desk calculator was D. A. Callin. The IBM operators ran into trouble with accuracy. In the meantime, Callin plugged along and got our first answer. So we started on the next case, and the IBM operators still had trouble, but Callin got the second answer, and so it continued until we finally calculated 26 different cases, 25 by Callin. The IBM machines finally gave us two answers, which only checked the hand calculations already completed. Each case required just over 100 h of continuous hand calculation, often involving double-precision arithmetic. You can see that the calculations took over a year of continuous work by one man. Callin could do all this effortlessly, often working two days without a mistake. If we had not been so fortunate as to have his help, I doubt if we would have been able to finish the problem, certainly not by the time we needed results. Thus, whenever I think of him I think of him as the virtuoso of the Marchant.

The calculations and work were successful, and by the spring of 1952, we had most of the answers to the Görtler stability question. At the same time we applied our results and found that an amplification ratio correlated data much better than Görtler's parameter $R_\theta\ (\theta/r)^{1/2}$ at transition. In fact we found that Görtler type transition occurred about when $lnA/A_0 = 10$, A/A_0 being the amplification ratio. This work was nearly complete in early 1952, but was all classified. The internal Douglas report, ES 17110, was published in March 1953. We later studied and correlated further data which ended up as my paper on Görtler stability, including the e^n method of predicting this kind of transition. As noted, transition was found to occur when lnA/A_0 was about 10. Thus, you see that the DESA-2 airfoil had much to do with development of the amplification ratio method, first stimulated by the Görtler problem. For the very short highly concave regions, it did not stand to reason that a high value of Görtler's parameter, acting for an exceedingly short time, should cause transition. Rate and total growth of the disturbance somehow must be involved. For the DESA-2 airfoil in the worst concave region, Görtler vortices grew across it by a ratio lnA/A_0 of 5 at $R_c = 60 \times 10^6$ so we thought we were safe.

Naturally we wondered just how rapidly the convex regions would damp out the Görtler vortices. We tried stability calculations several times for the convex regions, but never could get convergence to an eigenvalue.

The other considerable effort was on the question of Tollmien Schlichting stability. It seemed to me that amplification rate and ratio for this kind of problem also gave the most rational guide to transition, but at the time, the e^n method had not been fully developed. In fact, the thinking was something of a carry over from the Görtler problem. The prediction of transition was in a primitive state. As I remember, the best way of predicting transition was to use some value of R_θ as the criterion. We needed our guidance around 1950 or 1951, and R_θ was the only guide. Significant papers on the subject such as Gazley's and Granville's did not come out until 1953. Michel's paper appeared in 1951, but we did not know about it until after the DESA-2 was tested.

R_θ at transition was not then known as a function of the pressure gradient or Pohlhausen's Λ. From NACA airfoil tests, we of course knew there was, indeed, an effect of pressure gradient, and we wanted a method of analysis that at least took into account some of the changes in stability caused by it. Pretsch's charts were a great help on this question and on the airfoil design. I do not remember all our thinking at the time, but in the report, the airfoil is very nearly designed as if the e^9 criterion was applicable. That is, at $R_c = 50 \times 10^6$, the calculated amplification ratio for the entire airfoil is $e^{9.5}$ on the upper surface and $e^{7.2}$ on the lower surface. In fact, in the report,[2] the design conditions are stated to be as follows.

1) Airfoil thickness to be 6.6% in order to be usable on the F4D-l airplane, if desired.
2) Design lift coefficient = 0.1.
3) $C_{m.25}$ to be zero at $C_L = 0.1$.
4) At $R_c = 50 \times 10^6$ for Schlichting waves the amplification ratio at the trailing edge must be less than exp 10.
5) At $R_c = 50 \times 10^6$ the amplification ratio for Görtler vortices must be less than exp 10 and preferably as low as possible (for each concavity).
6) From the practical standpoint, the design should have as few slots as possible.

Thus, it is seen that the stimulus for the e^9 method lay in the DESA-2 airfoil. By 1952, a large number of amplification ratios at transition had been calculated from the various test data available, and while I believed they formed the most logical method of prediction, there was a large amount of scatter in the calculated values, so no rational method of prediction was evident. That was not so true for the Görtler problem. In it, for the available test data, the calculated amplification ratios at transition were fairly uniform, at a value $lnA/A_0 = 10$. Once, probably in 1953, I showed our studies of test data to a co-worker, Deane Morris, who later moved to the Rand Corporation. He saw the large scatter in calculated amplification ratios at transition and suggested that, instead of worrying about it, we just take some average value and see how well we could actually predict the transition point. We did and found the point was predicted well and rationally. Thus was born the e^n method, at least my contribution. After further studies and rounding out the work, I presented my paper about it in 1956. For test data at transition, the calculated amplification ratio is often a sensitive parameter. Conversely, the position of transition is relatively insensitive to the value of the amplification ratio.

It should be noted that this development was not started with the expectation of finding that an amplification factor alone would be a suitable criterion, because it is only something that multiplies an initial disturbance. Transition must depend on disturbance strength, not just a ratio. However, calculating the amplification ratio was a well- defined problem, and so it was done first. We then expected to study turbulence inputs at the neutral point. Some work was done, such as multiplying percent turbulence times amplification ratio to get a disturbance strength to see if we could predict the effect of turbulence on flat-plate transition. This proved no good; the problem is far more complicated. However, two things happened, the DESA-2 contract ended, and thanks to Deane Morris's suggestion, we found that A/A_0 by itself was a good parameter. That is about where the prediction of low speed transition stands to this very day.

A lesser development, but also stimulated by the peculiar pressure distribution of this airfoil, was my piecewise method of approximate laminar boundary-layer calculation.[3] The sawtooth pressure distribution gave me the idea that ficticious origins could be used for similar flows of the type $u_e = cx^m$, from which approximate solutions quickly could be patched together. We especially wanted to use similar solutions because Pretsch's stability charts were for various values of Hartree's β. As I remember it, this idea came up in some discussion one afternoon, and by the end of the day, we had made our first general calculation. A preliminary form is in the DESA-2 report.[2] We gradually assessed the method, refined it, and published it in 1956.

Another event in connection with the DESA-2 airfoil was an attempt at finite-difference boundary-layer calculation. To answer stability questions, we wanted to know very accurately the boundary-layer profiles between slots, and integral methods are inadequate. I studied various methods of writing a finite-difference system and selected one that seemed the best. I then gave it to the head of computing, W. C. Schlieser, and asked for an estimate of the machine time required to calculate 20 or so stations. He came back with an estimate of 2000 machine hours! That stopped my interest in finite-difference methods for the time being. The computer was the IBM Card Programmed Calculator—a semimechanical computer.

In tests on the DESA-2, we could get 100% laminar flow only to a Reynolds number of about 6x 10^6, so the attempt and my airfoil idea were unsuccessful. I never understood why until 1978. Not knowing any better, I had assumed that the much longer convex regions between the concave regions would damp out the Görtler vortices so much that, at each new concavity, the flow would essentially start out afresh; Tollmien-Schlichting waves behave this way. When they encounter a sufficiently favorable pressure gradient, they damp out. Besides, since the airfoil has significant thickness, there was more convexity than concavity. In 1978, Professor Eli Reshotko pointed out to me the principle of exchange of stabilities. According to this principle, when the flow encounters a concavity, Görtler vortices amplify; but when they reach the convex portion, they do not damp, they remain neutral in stability. Mathematically speaking, it works out that, for concave flows, the exponential giving the eigenvalue has both a real and imaginary part, the real part being a measure of the amplification rate. However, for convex flows, it is a pure imaginary, meaning an undamped fluctuation. Hence, the behavior of Görtler vortices is considerably different from that of Tollmien-Schlichting waves. In my original concept of the entire amplification process, I visualized it as a series of climbs and descents; but according to the principle of exchange of stabilities, the process would be something like a staircase: a climb, a level region, a climb, a level region, etc. So the total amplification ratio for the airfoil would be the product of the amplification ratios for each individual concavity. Also, this neutral stability for convex surfaces probably was the cause of our inability to converge to eigenvalues in our Görtler calculations when we tried such surfaces.

Therefore, the airfoil was not a practical success, but it was an indirect success because, with its peculiar sawtooth pressure distribution, it was a great stimulus to development of boundary-layer methods.

Another study was of the size of roughness that could trip a laminar boundary layer, done for the U. S. Navy. While our work on developing a laminar-flow airfoil had stopped, we still had our knowhow, and in 1954, the U.S. Navy asked us to look into the practical problems of laminar boundary-layer control. It seemed to me that one of the most severe practical problems was that of roughness tripping the boundary layer. We made a large wind tunnel model and very carefully and extensively looked into this problem. Our interest was the height of roughness that first affects a long laminar run. There had been many tests by others, e.g., Dryden, Tani, and Potter, on the roughness required to bring transition to the roughness but not on the height of roughness that first affects transition. About the time we had obtained all our data and were studying various ways of correlating it, I talked to Professor Hans Liepmann. He commented that he and Frank Goddard were having very good luck on the problem of turbulent skin friction of rough surfaces at high Mach numbers by correlating test data with really local conditions at the bottom of the boundary layer. We followed suit and obtained very successful correlation by using the true Reynolds number of the roughness, $u_k k/\upsilon_k$ where k is the height of the roughness, u_k the velocity in the boundary layer at the top of the roughness and υ_k the kinematic viscosity at the same point. Both on this problem and many others, a great many results were all explained by correlating with the true local flow at the roughness. Hence our advance was in the correlation method, and it was due a good deal to the conversation with Hans Liepmann.

The Navy did nothing more than ask us to look into the practical problems of laminar-flow control. Thus, the idea of tests of roughness were entirely our own, in fact, my own as I remember. Also, the idea of the DESA-2 project was entirely my own. There was no spirit of competition on this problem or trying to out-do others, except very indirectly as an improvement over Griffith airfoils. On both the transition and piecewise method of boundary-layer calculation, one might say we were working in a vacuum. The procedures just seemed quite logical, so we explored and implemented them.

3. Some Comments on the Practical Side of Getting Work Done

The two wind tunnel models were rather expensive, and a great deal of theoretical work was involved that took manhours. Experienced readers may wonder how I managed to get the work authorized. I do not really know because I did not have any problems like those in later years. For exploratory theoretical work there was a substantial budget of indircet time (called 9603 time at Douglas) that one was relatively free to charge to in the late 1940's. Gradually, due to tighter budgetary controls, this fund became smaller and smaller over the years.

I do not remember any specific battles to sell authorization of funds for the $G0010_7$ or DESA-2 airfoils although I do remember talking about them and their principles. The work was authorized as amendments to the XF4D-1 tailless airplane contract as peripheral studies with the idea that results might be appli-cable to the airplane in the future. So you see that funds did not have to be contributed by the company, they came from the Navy. I have always found the Navy alert to developments that might lead to better air-

Table 1. Summary of principal methods for solving the laminar boundary-layer equations

NO.	METHOD	WORKING FORM OF EQUATION	IMPORTANT CONTRIBUTORS	ACCURACY AS MEASURED BY ERRORS IN t, THE SKIN FRICTION	ADVANTAGES	CRITICISMS	PRINCIPAL REASONS FOR REJECTION AS BASIC METHOD FOR PRESENT INVESTIGATION
1.	Momentum Integral	Von Kármán's momentum integral equation	Von Kármán (Ref. 1) Pohlhausen (Ref. 1) Walz (Ref. 1)	3 figure accuracy possible in the most favorable cases. Greater than 100 percent error in adverse cases.	Very fast and convenient. In well conditioned problems accuracy is good*.	Accuracy entirely unsatisfactory for problems of boundary layer stability. Method insensitive to shape of upstream boundary layer. It considers only thickness.	Completely unable to handle many problems of importance. Poor, even on well conditioned problems in compressible flow.
2.	Correlation	None, it is based on empirical correlation of available exact solution.	Thwaites (Ref. 13)	About the same as above.	About the same as above.	About the same as above.	Completely unable to handle many problems of importance. Extensive compressible solutions essential for correlation are not now available.
3.	Local Similarity	Based on similar solutions, as of Falkner-Skan equation.	Smith (Ref. 16)	About the same as Nos. 1 and 2	About the same as Nos. 1 and 2.	About the same as Nos. 1 and 2.	Method has promise as an approximate procedure, but little work has been done. At best it will be completely unable to handle many problems of importance.
4.	Infinite Series	Various transformations of momentum equation where s, y remain essentially as the independent variables.	Blasius (Ref. 1) Howarth (Ref. 1) Tifford (Ref. 14) Görtler (Ref. 1) Cope and Hartree (Ref. 15)	5 to 8 figure accuracy possible in the most favorable cases, as near stagnation point. Several hundred percent error in adverse cases.	Exact solutions can be obtained easily for certain fortuitous edge velocity distributions.	Cannot handle ordinary pressure distributions such as those of airfoils.	Unable to handle the most common type of pressure distri-bution. In many high speed flow there are so many variables, that calculation of universal functions is impractical.
5.	Dorodnitzin	Transformed momentum equation in which x and y remain essentially as the independent variables	Dorodnitzin (Ref. 17) Pallone (Ref. 17)	Theoretically exact, but insufficient work has been done to establish accuracy-computing time relation. High accuracy will probably require lengthy computation	Principal advantage is ability to give much better answers than No. 1 with reasonable labor. Method is essentially a generalized Pohlhausen method.	For high accuracy a high order system of simultaneous first-order differential equations must be solved. Starting requires special procedure.	Computing and setup time appears unreasonably long when high accuracy is desired as for B.L.Stability investigations.

Table 1. Summary of principal methods for solving the laminar boundary-layer equations (cont.)

NO.	METHOD	WORKING FORM OF EQUATION	IMPORTANT CONTRIBUTORS	ACCURACY AS MEASURED BY ERRORS IN t, THE SKIN FRICTION	ADVANTAGES	CRITICISMS	PRINCIPAL REASONS FOR REJECTION AS BASIC METHOD FOR PRESENT INVESTIGATION
6.	Finite Difference, Explicit	Unmodified momentum and energy equations.	Wu (Ref. 18) Görtler (Ref. 1)	Theoretically exact, but insufficient work has been done to establish the accuracy, computing time relation.	Simple and sufficiently fast for low accuracy. Theoretically can handle any boundary layer problem exactly in limit.	Boundary layer thickness in this system varies greatly, causing computing complications. Computing time excessive where high accuracy demanded. Has numerical stability problem.	Accuracy-computing time relation is poor, if high accuracy demanded. Starting the solution is a problem. Numerical stability question is a disadvantage.
7.	Finite Difference, Explicit	Crocc's transformed equation, in which x, u are independent variables.	Baxter and Flugge-Lotz (Ref. 19) Raetz (Ref. 20)	Theoretically exact. Accuracy high (3 to 5 figure) Computing time is long.	Theoretically exact and convenient formulation, because boundaries are at u = 0, u = 1, and Crocco's equation is second order.	Computing time great where high accuracy demanded. Crocco's equation not suitable for problems involving "overshoot". Has numerical stability problem.	Computing time too long, "overshoot" difficulty eliminates Crocco's equation as basis for a completely general method.
8.	Finite Difference, Implicit	Crocc's transformed equation, in which x, u are independent variables.	Kramer and Lieberstein (Ref. 21)	Theoretically exact, but insufficient work has been done to establish accuracy-computing time relation. Computing time should be short.	Same as No. 7 above plus rapid computing time. Computing is fast because a "trick" good for second order equations has been used. No numerical stability problem.	Crocco's equation not suitable for problems involving "overshoot".	"Overshoot" difficulty eliminates Crocco's equation from consideration.
9.	Hartree-Womersley Finite Different Implicit	Simple transformation of momentum equation in which x, y remain essentially as independent variables.	Hartree (Refs. 4,5,6) Smith and Clutter	Theoretically exact. Accuracy high (3 to 5 figure). Computing time now medium, but could be reduced greatly.	Theoretically can handle any boundary-layer problem, exactly in limit. Computing time is good. Has best staging procedure of any exact method. No numerical stability problem.	Probably slower than No. 8 in cases where No. 8 can handle the problem.	

planes. Because the work was under amendments to a classified military contract, it too was classified until the end of 1955.

The computer mainly used at this time was the IBM Card Programmed Calculator (CPC). It was an electromechanical relay type and a great advance over a Marchant-type desk calculator, but almost infinitesimally slow compared with modern calculators. A set of IBM cards was properly punched, using machine language, and put in a hopper to instruct the machine on what to do. It was a considerable advance over earlier ones for which a plugboard had to be wired. Storage was limited and mechanical. Computing rate was governed by rate of feed of the card deck. I did not observe the operation of the CPC closely when we tried to solve the Görtler problem but I did on another problem.

The original table of solutions to the Falkner-Skan equation were calculated by D. R. Hartree, using a Bush-type mechanical differential analyzer which yielded between three- and four-place accuracy. In our airfoil and other work we used his solutions so extensively that it seemed to be a good idea to work out more complete and more accurate solutions. Now that a much better and digital type of computer was available, it did not seem to be a large job.

Accordingly, we started working, charging to the general 9603 indirect charge number. (At Douglas, computer use was free just like a desk calculator, to foster their use; so in the good old days the only problem was access.) In a moment of optimism I decided to make tables of five-place accuracy. Only later did I gradually realize that developing that accuracy for certain on a nonlinear equation is a formidable undertaking. We solved the equation by a Taylor series extrapolation, beginning the work at Santa Monica because El Segundo's computer was not yet operable. After a while the work was transferred to El Segundo, where I often ran the machine myself, using the shooting method to meet boundary conditions.

If about 100 steps were taken and with the proper number of terms in the series to assure five-place accuracy, a run with given initial conditions would require about 15 min. One looked at the tab sheet to see whether the boundary-layer profile went above or below the asymptote. Then a new initial condition was punched into a card and a new try made. If one had uninterrupted use of the machine, a proper solution for a particular value of Hartree's β could be made in about half a day. So when I was doing this work I was mostly standing around, watching the tab sheet being printed, making plots to estimate the new value of Hartree's F_W, etc. I hope this story gives you an idea of what automatic computing was like in its early days. The tables were published by the Institute of Aeronautical Sciences as a Sherman Fairchild Fund paper in 1954 but all the work was done in 1951.

One other aspect should be mentioned. When the program was transferred to El Segundo from Santa Monica, I naturally duplicated some runs. The printout was eight-decimal places, I believe. For a number of steps the new and old tab sheets would check exactly. But then after a while there would be a gradual drift; first the eighth place would not check, then the seventh, and finally down to the sixth or fifth, so that I could not be certain I met my goal of five place accuracy. I talked and worked with the computer staff on the problem for about two weeks and no explanation was found. The discrepancy could not be tolerated, so in desperation I sat down at a Marchant and, using exactly the same mathematics and roundoff procedures, set out to duplicate a run by hand. After about a day and a half of steady calculating I found my first disagreement. This gave some guidance as to where to look for the trouble. The deck of cards containing the program for this problem was about 0.9-in. thick. In order to reduce the frequency of reloading the deck in the hopper, eleven duplicate decks were made, making a dozen in all. What we found when we traced through all the decks was that one card had been misplaced so that eleven times instructions were correct but were wrong the twelfth time. The erroneous card determined the value of one of the last terms in the Taylor series extrapolation so the error was small. This experience has made me extremely cautious about trusting the output of a large scale computer on a complicated problem, because there are so many possibilities for error.

I was not sure of the propagation and growth of roundoff errors, so after finding all the solutions, I checked by rerunning them again with two more terms in the Taylor series and longer steps. Everything checked. I hope this chronicle gives you the flavor of automatic computers in the early days.

4. Developing a General Method for Calculating Laminar Boundary Layers

After my success with the panel method of solving the Neumann potential flow problem, I chose as a home study project an attempt to solve the inviscid transonic flow problem. I studied at home for about a year without success. I considered all kinds of methods, but did not consider pure finite-difference calculations because the computing machines were too slow.

Thus you see that my sole interest was not boundary layers, but after about a year of frustration, I turned back to them. Now my goal was not to control the boundary layer but just to try to tell what it would do on its own. I wanted a method that would be accurate enough for stability calculations and that would be quite general. At the time, some methods had accuracy but were not sufficiently general, and other methods had generality but were not sufficiently accurate. In fact, Table I, taken from report ES 40446,[4] is an excellent summary of the methods available at the time and their limitations. All, except the last method which was our development, had some sort of limitation. It should be said that Oscar Seidmann of the U. S. Navy supported our work, but he questioned whether our proposed work was any advance, and for this reason, the rather elaborate chart was prepared.

In trying to find a truly general and accurate method, I had no idea what to do. I read the literature extensively, both at Douglas and at home. I studied various forms of equations. I studied various methods of solution. There are many types of transformations, Crocco's, von Mises', Falkner and Skan's, Görtler's, Cope and Hartree's, and many others, not to mention all sorts of integral methods. For most transformations, I found some drawback. After looking at these various forms for some time hoping to find something that would give an easy solution but not finding it, common sense got the better of me, and I realized that what I wanted was in the full partial-differential Falkner-Skan equation. It did not distort the x axis and had no pathological problems such as with overshoot profiles, or starting. (An overshoot profile is one in which the velocity ratio u/u_e may exceed unity before decaying to its final asymptotic value. Accelerating flows past hot walls can generate them.) We gradually focused on this equation. but for a considerable time, we considered Howarth- or Görtler-type universal solutions for a series representation of the edge velocity. We studied this at Douglas. Both velocity and temperature distribution were represented by series expansion but I do not remember all the conditions being represented. Hajimu Ogawa, now a professor of mathematics, worked on this problem, and he obtained universal solutions that were far more general than any in existence, such as those of Tifford. It was a major accomplishment and easily could have been an important paper.

However, I got to thinking. To have an adequate supply of universal solutions for the many variables of a general problem, a very extensive library of universal solutions would be needed. I recalled the old saying, a table of one variable requires a page, two variables a book, three an encyclopedia, and four a large library. I realized that the number of differential equations that must be solved to provide the necessary universal functions would be far more than if I solved the basic Falkner-Skan partial differential equation directly for each indivi-dual problem that came up—at least in my lifetime. Besides, the universal-series-type solutions are poor for airfoil-type pressure distributions. Therefore, after accomplishing all this good work, we dropped it cold.

Because I knew Galerkin's method very well from the Görtler stability problem, we tried this form of solution, as far as writing out the system of simultaneous equations; but it was nonlinear, and we did not like it. Another that we studied was Picard's method, the method of successive substitutions. We tried it on Hartree's ordinary differential equation, and it was fast and converged beautifully for all values of β. I have already mentioned that I solved this same equation some years before by the shooting method, to construct the tables of solutions of the Falkner-Skan equation for various values of β.

With our success using Picard's method, we obtained Navy support to try to solve the full partial-differential equation. To do that, we planned to represent the x-derivative terms by finite-difference approximations which then gave us an ordinary differential equation at such x station to solve in the y direction. Hartree and Hartree and Womersley had used this method successfully with the old Bush mechanical differential analyzer to solve a number of problems.

Hence, in getting under contract with the Navy to solve the full equations, we had confidence because of our success with Hartree's ordinary differential equation. The additional x derivative terms looked small, and we did not see them as any problem. In fact, we considered the flows more or less as locally similar, and the x term would merely be a slight correction.

However, that is not the way it worked. Picard's method now would not converge, and try as we might with various alterations, it still would not converge. The contract time was running out, and we were getting desperate. So one day, we dropped Picard's method and switched to the old shooting method that had been used without trouble on the Hartree β profiles. It worked fine. That is how our original method was born. It was hardly a cold, deliberate, and logical study that then gave birth to a successful method.

The original work was on the momentum equation alone. It is the worst problem, because the equation is nonlinear. After success with it, it was fairly easy to extend the same method to include the energy equation. Further extension was made to handle the species equation for such things as nonequilibrium flows. A summary of this work is in a Pergamon Press publication.[5] This method could truly produce high-accuracy answers for any two-dimensional or axisymmetric problem that arose. Any *Pr(x,y)* (Prandtl number), any edge velocity, any suction distribution, any wall temperature distribution, or any heat transfer distribution could be analyzed. The only restriction we encountered was a sensitivity to short x steps toward the rear of the body, but that was overcome. Our original Douglas report was issued in July 1961.[4] While the method has been largely superseded, it still remains as one that is both accurate and simple to write out and program.

5. Other Laminar Work

About 1965, Professor Marten Landahl and R. E. Kaplan came out with their purification scheme of solving the OrrSommerfeld equation. Because this broke a bottleneck in the numerical solution, we decided to get active in this area and I talked it over with R. S. Shevell, Director of Aerodynamics, pointing out the gains in knowledge and the possibility of getting research-type contracts. He approved getting into the field; then I proposed work on it to Professor A. R. Wazzan, a consultant. He studied and learned the problem and the field and began calculations. Shortly after we began our work, a new book by Bellman and Kalaba, entitled *Quasilinearization and Nonlinear Boundary Value Problems,* came out in 1965. Bellman and Kalaba had very good reputations for contributing useful methods, so I bought the book and gave it to Tom Okamura in my group to see if it had anything useful in it. He discovered Gram-Schmidt orthogonalization which was a more rigorous and powerful generalization of Landahl and Kaplan's purification technique. I had nothing to do directly with its appli-cation but do claim enough vision to buy Bellman and Kalaba's book. I do not know how we stand timewise with respect to the orthogonalization procedure and other workers. For instance, Radbill et al. came out with the method at about the same time, but at least we discovered it independently. Eigenvalues are basically found numerically by calculating independent solutions of the Orr-Sommerfeld equation and then combining them to meet boundary conditions. One of the solutions used is a weakly growing exponential and the other is strongly growing. Due to roundoff the strongly growing solution can find its way into the weakly growing solution. Then, at high Reynolds numbers after calculating for some distance in y, one does not have the necessary two indepen-dent solutions needed for meeting boundary conditions. They are proportional. Gram-Schmidt orthogonalization keeps them independent.

After learning and exploring the Gram-Schmidt orthogonalization procedure which now made it possible to reach almost any Reynolds number, I thought a very useful contribution would be to modernize Pretsch's old Falkner-Skan stability charts and do both spatial and temporal amplification calculations. We did this, and the charts were first published in 1968.

The original e^9 method used Pretsch's charts, hand calculation, and interpolation between charts. We now had the capability of solving the stability equation at each x station for the exact profile at that station, supplied by the exact solution of the partial-differential boundary-layer equations as described earlier. Okamura, Jaffe, and Wazzan carried out this work which became a general and highly accurate method of calculating amplification ratios, frequencies, etc. The method then became rather dormant at Douglas Aircraft, because transports have no laminar flow.

Around 1974, the Navy first found out about this work and that it could analyze both two-dimensional and axisymmetric flows, the effect of heating and cooling in both water and air, suction, etc. Because such calculations were of little direct interest to Douglas, I proposed to the Navy that we pull together our potential flow calculations that gave pressures, our boundary-layer calculations that gave boundary-layer profiles, and our stability calculations that gave amplification rates. They bought the proposal, and it was pulled together by A. E. Gentry with A. R. Wazzan handling the stability aspect. Gentry named the program TAPS for transition analysis program system. It is now in widespread use by various Navy organizations doing preliminary design studies of torpedoes and other bodies.

One of the reasons the stability program was of interest to the Navy was that it had been extended to handle the problem of water flow past warm walls. This modification began in the middle sixties while we were still developing our boundary-layer and stability methods. I was aware that accurate solutions of boundary-layer flows in water with heat transfer did not exist so I suggested them as a kind of homework exercise for developing our methods, Douglas having no basic interest in water flows. These were worked out and Professor Wazzan made a simple but accurate modification of the Orr-Sommerfeld equation to analyze their stability. Both the boundary-layer calculations and the stability analysis were presented in 1967 at an ASME meeting. The stability analysis revealed a profound effect, a very small amount of warming the walls increases the stability very, very much, almost as much as suction. At about the same time (1968), Hauptmann discovered the effect, using approximate methods, whereas our methods were highly accurate. We had no knowledge beforehand of Hauptmann's work.

6. Turbulent Boundary Layers

At least for airplanes, laminar boundary layers are not the common case, turbulent layers are. Recognizing this fact and now having success with the laminar case, I began to think about the turbulent problem, not in general, but only for boundary layers. It seemed to me there were certain basic relations that applied regardless of the pressure distribution—the law of the wall, Prandtl's mixing length relation, the velocity defect law, intermittency at the edge of the boundary layer, and Van Driest's modification of the law of the wall. Previous attempts at calculating turbulent boundary layers had all used approximate forms of Prandtl's equations or even none at all. I wondered what kind of answers we might get if we incorporated these rather universal laws into the exact equations now being solved regularly for laminar flows. Incorporation as we did it just meant adding an algebraic turbulent eddy-viscosity term to the equations. Of course it was not a constant.

More specifically, after examining the available literature and information on the subject, such as Townsend's and Hinze's books, I used an inner eddy-viscosity formula, an outer eddy-viscosity formula, and a decay law. For the inner eddy-viscosity formula, Prandtl's mixing length formula basically was satisfactory, but I liked Van Driest's improvement of it, which had a certain amount of physical justification. The formula used was

$$\varepsilon_i = \rho\kappa^2 y^2 \, (1 - e^{-y/A})^2 \, \frac{\partial u}{\partial y}$$

Here κ is von Kármán's constant. The $\rho\kappa^2 y^2$ quantity is just Prandtl's mixing-length formula. Van Driest's modification brings solutions properly to the very wall. For the outer region, we used a form suggested by F. Clauser, but with a slightly different constant, namely,

$$\varepsilon_o = 0.0168 \rho u_e \, \delta^* \gamma$$

As you can see, at any particular station ε_o is a constant viscosity except for γ, the intermittency factor. Finally, as one gets to the outer edge, intermittency sets in and the turbulent transport processes fade away. We used Klebanoff's expression for γ to describe this fadeout. So you see that, at the very wall, the turbulent viscosity was zero, but then it grew mainly as y^2, but as y^4 at the very wall, which is theoretically correct. Then when it reached the value indicated by the outer formula, we would switch to it and finally fade everything out as the edge was reached. This formulation is quite simple but the results were in reasonable agreement with the experimental data that was available. While numerous extensions of the method have been made, it is interesting that my original formulation has never been changed, not even the constants.

It is interesting to note that I often received a lot of "static" from 0. R. Dunn, a later Director of Aerodynamics, about our laminar efforts. He would ask, "Why are you wasting your time on laminar boundary layers when the DC-8 is all turbulent?" One day in connection with this turbulent extension when he again asked me this question, I went to his blackboard and wrote out the full incompressible equation including the turbulent viscosity term ε. He could see that the structure was the same for both the turbulent and laminar cases, and that work on the laminar was a kind of training exercise for work on the turbulent case. After that, he became a big booster for all of it.

I did not have a high degree of confidence that incorporating accepted basic turbulent properties in the full equations of motion would give good answers for all kinds of boundary-layer flows because an algebraic eddy-viscosity model has obvious flaws. Instead, initially I regarded the effort as one merely to find an answer to the question, "If we use the best available algebraic transport properties plus the exact equations of motion, how good are the answers?" Our first efforts were just to use the described eddy-viscosity formulation and introduce it into the equations of our shooting method. We then solved a number of cases by this method to see what kind of accuracy was indicated and found it to be very good for a considerable variety of problems. Our first paper about this method was given in 1966.[6]

The studies seemed to answer my question favorably, so we went ahead with development. The shooting method was very slow; the eddy viscosity was a variable function of the boundary layer itself, so a double iteration was needed. Hence I started out to find a better method. After reading Fox's book, *The Numerical Solution of Two-Point Boundary Problems*, I settled on a five-point finite-difference method. We were developing and applying this method at the time T. Cebeci joined my group at Douglas. He soon got into this work and made the new finite-difference method operational. For several years this appeared to be the leading all-around method for boundary layers. Dr. Cebeci devised several ingenious extensions to the eddy-viscosity formulations to account better for suction and blowing, low Reynolds number, heat transfer, etc. This extended version did very well at the Stanford Symposium in 1968.

The method is still used throughout the country but, it has been considerably superseded by Cebeci's extension to Keller's Box Method, a powerful and more universal numerical method than my five-point method. This is only a mathematical method of solution; the equations are just the same. There now are numerous turbulent methods, some far more sophisticated than mine. Sometimes their predictions are better and sometimes worse, so on the average our method is still holding up for ordinary boundary layers. When Cebeci and I wrote our book about turbulent boundary layers, I feared it might become obsolete rather fast because there was great activity in developing various transport equation methods. They definitely are more logical, but still lack a basic ingredient—a sounder description of turbulence—and so seem rather bogged down. For ordinary boundary layers, our simple algebraic eddy-viscosity formulation seems as good as any and much easier to calculate. Ordinary boundary layers are the important kind on aircraft.

When I began to try to calculate the turbulent boundary layer, I was not aware of any similar effort. More accurate calculation of turbulent boundary layers was just a desirable goal. After a while, I became aware of Mellor's work which used a similar approach, and of Bradshaw and Ferriss's work, but not at first, when the course of development was being set.

7. Closure

A final contribution in the field of turbulent boundary- layer theory was my extension to the axisymmetric case of Stratford's method of predicting separation. Stratford's method is an approximate procedure that is easy to use and reasonably accurate and so is very useful. At the time I did this work, the main methods of predicting this kind of separation were by Presz and Pitkin or else by some method like the Cebeci-Smith that involves the full partial-differential equations. These are rather long, and I wondered if one could not find a shorter method—an extension of Stratford's. I was at UCLA at that time and had some free time before a research contract on torpedo-like bodies came in. So I decided to use this time by seeing if I could work out the extension. I managed to. Stratford's method involves a transformation of boundary layers from one position to another. It is interesting that the old DESA-2 airfoil report materially helped me on this problem because, in it, I had worked out a related transformation for the boundary layer on each side of a suction slot. A paper about the Stratford extension was published in

ZAMP and dedicated to Professor Nicholas Rott on his 60th birthday.

In thinking back about my contributions, I find a very high percentage of my really basic ideas were conceived at home, not at work. Generally I was not seriously thinking about a problem, but still it was in the back of my mind, and suddenly I saw a solution. Another process that is very common is as follows. I would work on some problem rather intensively all day long, without satisfaction. I would usually be trying to implement some line of action but without full success. During the day, the problem was studied thoroughly and learned in intimate detail. Then frequently, when I had quit for the day, given up thinking, and freed my mind from the course followed all day, I would suddenly see the satisfactory solution I had been searching for. I think my subconscious brain is smarter than my conscious one. Also, I am more or less a loner. I have never been especially good at theoretical work and have not been very good at theoretical exchanges and mutual creation of ideas. The only way that has been successful for me has been continuous and persistent study by myself in some area. I am rather slow to understand a subject and, consequently, have to work on it very hard. However, when I finally understand it, I understand it well.

While working on research within the engineering department of an aircraft manufacturing company (Douglas Aircraft Co.) has its disadvantages, there are certain advantages. One is that a person is more intimately exposed day to day to real life problems and directly feels the great pressure for better and practical answers. This fact was often in the background of my contributions. I would see some area where there were problems, then gradually, I would see the essence of the problem, e.g., general turbulent boundary-layer calculations. Then my group and I would figuratively go off to the side and work on a fundamental solution. Being a research group, we could take a longer range approach than the design engineer who probably needed an answer the next day.

In general, at least for the most significant or important problems, I had little trouble getting the support of management. Of course I had to show clearly the goals and gains that would occur if a project was successful. We were rather selective about seeking government support, going after it only when we thought we could accomplish the job and at the same time benefit the company. This is why we solicited considerable work from the Navy in hydrodynamics, because much hydrodynamics is just incompressible aerodynamics. Also, I carefully watched another aspect—to be sure our work was useful to the company. I have seen a number of industrial research organizations founder, probably because their work was too far out or too unrelated to the company's more immediate problems. So we only worked on problems for which we could make a strong defense if attacked, the Neumann program, for example. I had a simple motto "there are far more problems than we can possibly solve, so we might as well pick on ones that are useful."

One final comment is in order about why I gradually drifted into research. I suppose the ambition of most young engineers is to be a chief engineer. It was one of mine and I attained it very early—at Aerojet. However, I soon found it was not what I wanted. I found myself just interpreting other people's input which at times I believed was biased. Gradually I realized I preferred creating information rather than interpreting something given to me by others. This is just a way of saying I preferred the technical side or more specifically creating information, which is just research. So after going back to Douglas I stuck to the strictly technical.

I hope this confession is of interest and helpful to others. Also, I would like to make it clear that if I had not had a group of very capable people at Douglas to help me implement the methods, much less would have been done. To them, who are too many to name, I feel much gratitude; I was lucky to have such good help.

Acknowledgment is made, with thanks to Richard S. Lee for suggesting the title.

References

[1]Schlichting, H. and Bussmann, K., "Exakte Lösungen für die laminare Grenzschicht mit Absaugung und Ausblasen," Schriften der Deutschen Akademie der Luftfahrtforschung. May 7, 1943.

[2]Smith, A. M. O., "Design of the DESA-2 Airfoil," Douglas Aircraft Co., Rept. E5171 17, AD 143008, Nov. 1952.

[3]Smith, A. M. O., "Rapid Laminar Boundary Layer Calculations by Piecewise Application of Similar Solutions," Journal of the Aerospace Sciences, vol. 23, Oct. 1956, pp. 901–912.

[4]Smith, A. M. O. and Clutter, D. W., "Solution of the Incompressible Laminar Boundary Layer Equations," Douglas Aircraft Co., Rept. ES 40446, AD 266271, July 1961.

[5]Jaffe, N. A. and Smith, A. M. O., "Calculation of Laminar Boundary Layers by Means of a Differential Difference Method," Progress in Aerospace Sciences, Pergamon Press, 1972, pp. 49–211.

[6]Smith, A. M. O., Jaffe, N. A., and Lind. R. C.. "Progress in Solving the Full Equations of Motion of a Compressible Turbulent Boundary Layer," presented at the Seventh BOWACA Symposium on Aeroballistics, U. S. Naval Missile Center, Point Mugu, Calif., AD634988, June 1966.

AMERICAN
INSTITUTE OF
AERONAUTICS AND
ASTRONAUTICS

WRIGHT BROTHERS LECTURE

"Commemorates the first powered flights made by Orville and Wilbur Wright in 1903"

given by

A.M.O. Smith

at the
6th Aircraft Design, Flight Test and Operations Meeting
Los Angeles, California

"High Lift Aerodynamics"

August 12, 1974

President

Chairman Honors and Awards Committee

AIAA Wright Brother's Lecture Award, 1974

6

High Lift Aerodynamics

by
Wayne Oliver
A member of AMO's Research Group

AMOs' exposure to high lift aerodynamics began early in his aeronautical career as detailed in his paper "High-Lift Aerodynamics" (ref. 1). He describes his M.S. thesis project, a boundary layer controlled straight wing design with slot Suction at 70% chord. One of the points made about this test was the lack of analysis capability in 1938 for multiple airfoils and turbulent boundary layers on arbitrary pressure distributions. Over the length of his career AMO and his Aero Research group would change that situation.

He was introduced to the sensitivity of stalling characteristics of wings to the boundary layer state near the suction peak by a change in paint scheme! The XBT-1 a forerunner to the famous SBD Dauntless alrcraft had good stall characteristics. The XBT-2 had exactly the same wing but had an unacceptable stall. The XBT-1 wing was completely painted with aluminum colored paint whereas the XBT-2 wing was painted yellow on the top and aluminum on the bottom. Where the colors joined at the leading edge, a slight ridge was formed by the masking tape during painting. John Wheatly, AMOs' boss, took out his pocket knife and scraped away the ridge. When the XBT-2 was stalled again, the vicious stall had completely disappeared. This interaction of the boundary layer with stall characteristics obviously made an impression on AMO.

After World War II AMO went into the aerodynamic research area and his efforts resulted in the Panel Method with successful calculations in 1955. Extensions of the method to lifting and multi-element lifting shapes followed over the years. These extensions are discussed in the section written by John Hess.

Another area of research was on the calculation of laminar and turbulent boundary layers on arbitrary pressure gradients. This research was accomplished mainly with and by Tuncer Cebeci and detailed in the section written by him. Transition was also studied with good success (e^9 method) and is discussed in the section written by Kalle Kaups. The point to be made, is that AMO contributed significantly to the generation of tools to evaluate practical aerodynamic problems. The high lift case is just a subset of the general capability of the methods. In addition to these important aerodynamic analysis tools, his major contribution in the high lift area would be the result of the AIAA asking AMO to write a paper on this subject in 1971. In 1974 he gave this paper "High-Lift Aerodynamics" as the Wright Brothers Lecture. It is a classic paper, and was given by AMO some 20 times including in Shanghai. Various parts of this paper were presented earlier in References 53, 56, 58 and 91.

AMO in the 1974 paper presents the "whys" and basic principles of high lift aerodynamics. As he noted in this paper that at the time of publication there were many erroneous theories in this subject matter. Some like the "blowing action of a slot" were still in text books and technical literature. His paper traces the history of high lift aerodynamics and notes those papers he considers significant and why. This gives the reader some perspective on what had been previously studied. Both slats and flaps are discussed. Limiting values of circulation are presented to show the boundaries of what can be expected both from potential and boundary layer flow. The calculation of separation location builds confidence in the tools that will be used to study the interaction of the boundary layer and arbitrary pressure gradients on separation location. This is followed by a detailed discussion of the canonical form for the velocity distribution and a detailed explanation of the advantages this form brings to the airfoil boundary layer analysis. Specifically, pressure rise shape and starting condition before the pressure rise on separation location are discussed at length. With this background, the paper then presents limiting cases, Stratford pressure recoveries, Liebeck airfoils, all neatly tied together based on the canonical form.

AMOs' paper (ref. 56) continues with, in my opinion, the most significant sections. He explains the five primary effects of gaps, or properly designed aerodynamic slots. The five effects are: Slat, Circulation, Dumping Velocity, Off-the-Surface Pressure Recovery, and the Fresh Boundary Layer

Effect. The discussion of these various effects really give the fundamental explanation of why multi-element airfoils can develop high circulation values. Never before had any author presented these principals together to form a unified picture of high lift aerodynamics. AMO deftly uses calculations by the potential flow and boundary layer computer programs to more clearly illustrate these concepts and typical design practice at the time. These sections alone make this paper required reading for anyone entering the specialty of high lift aerodynamics.

For completeness the paper includes AMOs' perspective on powered lift within the framework of the canonical form, tying previous concepts for unpowered systems to the powered cases (i.e., area suction, tangential blowing). Power profile airfoil airfoils are introduced which are thick airfoils with blowing at the trailing edge utilizing the Coanda effect and a movable plug nozzle. The transonic aspects of this type of design were only evaluated theoretically. Trailing edge blowing utilizing the Coanda effect on more conventional airfoils was successfully tested by Englar and flown on a modified A-6 Intruder aircraft.

Inverse methods for multi-element design are discussed. At the time of publications this research was just beginning, but two examples of results with existing methods are presented.

The paper concludes with AMO's thoughts on the future for high lift aerodynamics and the 10 most significant problems. Almost 24 years have passed since this paper and it is interesting to see which of these items are still in need of development. The Navier Stokes methods have helped in many of these areas if enough time and resources are allocated. In the high lift analysis techniques, drag, maximum lift and the stall type are still problems of significance for both two and three dimensional flows. Three dimensional flows are more complicated and are in a lower state of development.

AMO, with this one classic paper, gave the rest of us in the field of high lift aerodynamics a wonderful overview and explanation of the fundamental principals of this important area of aerodynamic design.

AMO at play

HIGH-LIFT AERODYNAMICS*

by
A. M. O. Smith†

c = chord
c_{dp} = section profile drag coefficient
c_f = local skin-friction coefficient $\tau_w/(1/2)\rho u_e^2$
c_l = section lift coefficient
C_L = lift coefficient
C_p = conventional pressure coefficient $(p - p_\infty)/(1/2)\rho u_\infty^2$
C_p^* = pressure coefficient when local flow is sonic
$\bar{C}_p$ = canonical pressure coefficient $(p-p_o)/(1/2)\rho u_o^2$
C_Q = suction quantity coefficient $Q/u_o x$
C_μ = blowing momentum coefficient $(2u_j^2 t/u_\infty^2 c)$, incompressible flow
$C_{\mu\theta}$ = blowing momentum coefficient referred to momentum thickness of the boundary layer at blowing location, $u_j^2 t/u_e^2\theta$, incompressible flow
f = chord fraction, see Eq. (5.1)
H = shape factor of the boundary layer, δ^*/θ
l = plate length
L = lift
m = exponent in $\bar{C}_p = x^m$ flows, also lift magnification factor (5.1)
M = Mach number
p = pressure
q = dynamic pressure
Q = flow rate
R = Reynolds number ($=u_o x/\nu$ in Stratford flows)
R_θ = Reynolds number based on momentum thickness $u_e\theta/\nu$
S = Stratford's separation constant (4.10); also peripheral distance around a body or wing area
t = blowing slot gap, also thickness ratio of a body
u = velocity in x-direction
u_o = initial velocity at start of deceleration in canonical and Stratford flows
v = velocity normal to the wall
V = a general velocity
x = length in flow direction, or around surface of a body measured from stagnation point if used in connection with boundary-lay flow

* 37th Wright Brothers Lecture. Presented as paper 74-939 at the AIAA 6th Aircraft Design, Flight Test and Operations Meeting, Los Angeles, Calif., August 12-14, 1974; submitted August 14, 1974; revision received December 27, 1974.

† McDonnell Douglas Corporation, Long Beach, California

Greek

α	=	angle of attack
δ	=	flap deflection
δ^*	=	displacement thickness of the boundary layer
θ	=	momentum thickness of the boundary layer
υ	=	kinematic viscosity
ρ	=	mass density
τ	=	shear stress
ψ	=	stream function

Subscripts

e	=	edge conditions
j	=	jet
l	=	lower
o	=	reference conditions, as in Stratford flows
u	=	upper
w	=	at the wall
∞	=	reference condition at infinity

1. Introduction and Purpose

When I first began work for the Douglas Aircraft Company in 1938, Donald W. Douglas, Jr., was in school. Several times, in connection with his studies, he came to me for help with aerodynamic homework problems, so that we got to know each other. Years later he gave me a 25-year service pin and asked me what I was doing. For once I was quick-witted and answered "still trying to understand aerodynamics."

That is the primary purpose of my lecture-the understanding of one aspect of aerodynamics. It is not a "how to do it" lecture, but rather one concerned with "whys" and principles. Since I am in research, I am not directly involved with the down-to-earth design problems of aerodynamic hardware, but instead am in a position of giving help to others. That reminds me of the definition of a mathematician I once heard. A mathematician is one who can tell you all about how to solve a problem but cannot actually do it himself.

It was my privilege as a graduate student at Caltech to listen to the brilliant beginnings of this series, by B. Melville Jones in 1937. It is a great honor, a privilege, and a difficult challenge to follow him and his many illustrious successors. In his introduction he said "...I am instructed that the Wright Brothers' Lecture should deal with subjects upon which the lecturer is engaged at the time, rather than with a general survey of some wide branch of aeronautical knowledge." To a great extent, that is my plan; the central interest will be subjects with which I have been closely associated.

Because I have been in the aeronautical field for some time now, I find it interesting to look back at the state of aeronautical knowlege when I was in school in the 1930's. I think you will see, as I proceed, that while the problems are far from being completely solved, our capabilities have advanced tremendously. My first problem of substance was my M.S. thesis project. It was to perform tests on a powered Boundary-Layer-Control Model in the GALCIT Ten-Foot Wind Tunnel. Figure 1 is a drawing of the model.

The project was inspired by German work of the time. In the design of the wing, the philosophy—so far as I knew—was as simple as this: 1) select conventional airfoils of the time for the wing, 2) put a slot (square edged) somewhere in the in the upper surface of the wing (70% chord seemed as good as

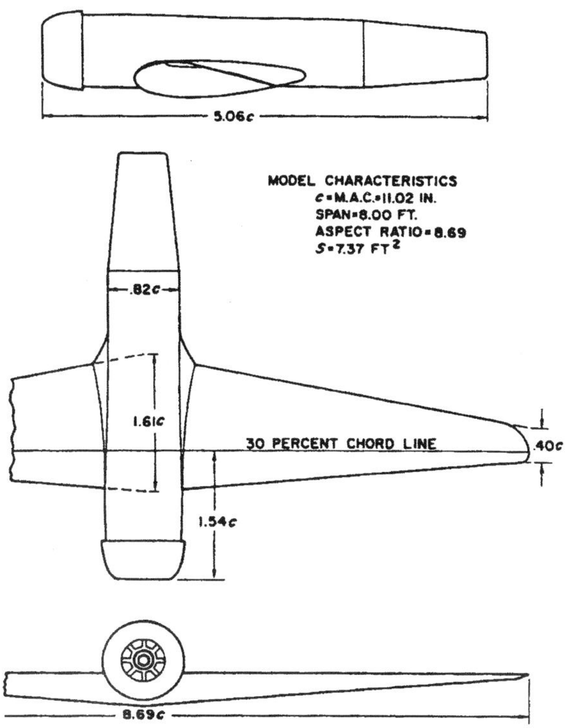

Figure 1. GALCIT boundary-layer-control wind-tunnel model.

anything), and 3) suck air and see what happened. There was insufficient knowledge of boundary-layer flow to do anything much more sophisticated. At the end of the tests, the writer had a vague uneasiness that there should be some more scientific approach to the problem of design. That is the way it was in a leading school that was under the leadership of the eminent von Karman and Clark B. Millikan.At that time, Pohlhausen's approximate method of calculating general laminar flows was a great new development. Only the most rudimentary method was available for estimating turbulent flows in general pressure gradients, for example, over an airfoil surface. The only convenient methods for analyzing inviscid flow about an airfoil were those of thin-airfoil theory and the flow about certain simple conformally mappable sections. Although Theodorsen's method had been developed, it was too tedious for practical use. To calculate the pressure distribution about an arbitrary airfoil amounted almost to a "stunt." Furthermore, all these examples were for single airfoils. Essentially nothing could be done for slotted airfoils. Hence it is no wonder that, in effect, people just drew up a shape by eye and tested it. Performance of airfoils was correlated in terms of certain obvious geometric parameters: camber, position of maximum camber, thickness, etc.

Transition as an explicit phenomenon in the development of a boundary layer was only vaguely recognized. For instance, at the GALCIT tunnel, surely as progressive as any, tests were often made on models of commercial and military airplanes to learn the effect of Reynolds number on drag. Drag coefficients were measured at a series of tunnel speeds. Results were plotted on a log scale and then generally extrapolated by a straight-line extension to full scale. The curve usually had a downward slope with Reynolds number, but not always. Even when the slope was positive, the line would be extended the same way; the effect was dismissed as a "poor Reynolds nurnber extrapolation." Now we know that the model must have been in the rapidly varying transition region, which must surely be left long before full-scale Reynolds numbers are reached. Fixing transition was essentially unheard of. B. Melville Jones, of course, greatly increased our awareness of the transition phenomenon.

2. Some History

Because, unlike the birth of Venus, new ideas do not burst forth fully matured or fully recognized, or even in one place, authoritative establishment of history is difficult. We know better than to attempt it here. Nevertheless, some review of highlights seems desirable.

In a search of the older literature, one of the strongest impressions gained is that in a sense there is "nothing new under the sun." Nearly all of the basic principles for influencing a flow to develop high lift have been known from the very early days of the airplane. The defects in knowledge were two; first, although what to do may have been known, the reasons for doing it were only dimly understood; second, quantitative analysis of a flow could rarely be accomplished.

Prandtl [1] had already conceived and demonstrated the principles of suction boundary-layer control in 1904, and by 1913, according to Weyl,[2] the notion of blowing to control the boundary layer had been advanced. One concept the author believes to be new is that of the jet flap, which seems to have been conceived and developed in the 1950's. Hagerdorn and Ruden [3] tested forms of the jet flap in 1938, but they did not understand what they were finding.

The ancestry of flaps can be traced back to the early days of flying. British R&M No.110, dated 1914 [4] contains one section entitled, "Experiments on an Aerofoil Having a Hinged Rear Portion." Figure 2 shows the shape tested. A large number of flap settings were examined. According to Weyl,[2] variable camber had been used even earlier. The LeBlon monoplane had a variable-camber wing formed by an adjustable part of the trailing edge. It was exhibited at the London Olympia Show in March, 1910.

Fig. 2 RAF 9 airfoil with a 0.385c plain flap tested in 1912-1913.

The idea of slats and the knowledge of the effectiveness of slots are nearly as old. In an important lecture given before the British Royal Aeronautical Society on February 17, 1921, Handley Page[5] described ten years of work on the development of airfoils that had slots. His work dealt not only with a single slot or slat but also with multiple slots. Figure 3 shows one of his models, which by current standards would be considered a relatively modern configuration.[6] Page 425 of [7] shows a good photograph of a 1920 airplane fitted out with slats.

Later in this paper, the author attempts to prove that an airfoil having $n + 1$ elements can develop more lift than one having n elements. Handley Page investigated this problem, up through 8 elements[5] Figure 4 shows one of his extreme airfoils, a very highly modified RAF 19 section, positioned at the angle for maximum lift.

Figure 5 shows lift coefficient vs angle of attack for the experimental airfoil

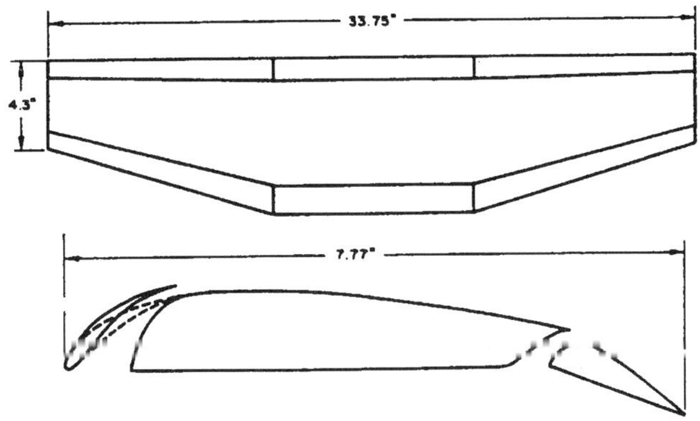

Figure 3 A wing tested by Handley Page as part of his effort at developing slots and slats.

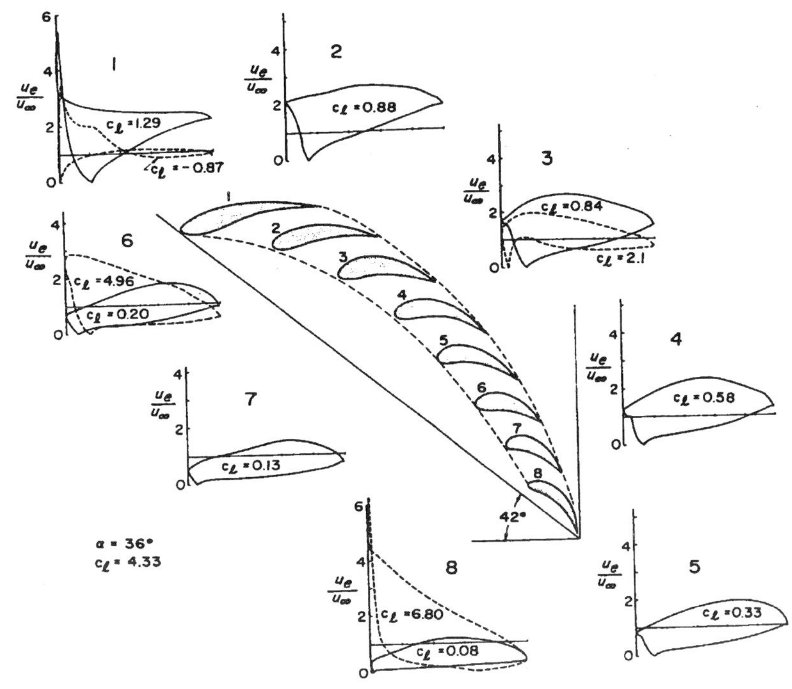

Figure 4 Handley Page's eight-element airfoil modified from an RAF 19 section. The model is at 42° angle of attack, the angle for maximum lift. Pressure distributions are theoretical. They were made at $\alpha = 36°$ to correspond to local angle of attack of the AR = 6 wind tunnel model. Theoretical c_l of ensemble is 4.33.

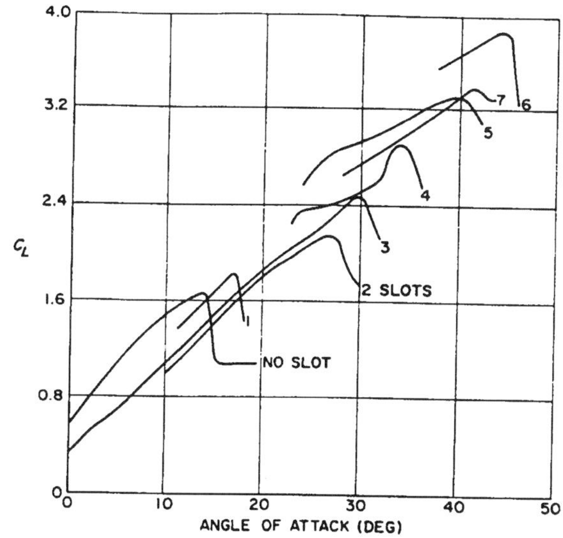

Figure 5 C_L vs α data for the RAF 19 broken up into different numbers of elements, as indicated by number of slots.

as it was modified from one to eight elements. It generally shows that the greater the number of elements the greater the lift; and it seems to confirm the author's deduction, which was made three years ago, quite in ignorance of these tests. The seven-element airfoil reached a lift coefficient of 3.92. Tests were made at a chord Reynolds number of about 250,000 on a wing of 6-in. chord and 36-in. span. Handley Page appears to have followed an empirical approach in his efforts. Concurrently Lachmann[7] at Göttingen was studying the problem theoretically. Lachmann used conformal-transformation methods and represented a slat by vortices in front of a circular cylinder. His work established the gross features and interaction effects for a slat and the main airfoil. Later, Lachmann and Handley Page joined forces. Further basic understanding and appreciation of the beneficial effects of a slat was gained by Le Page,[8] who systematically investigated the forces on two airfoils placed in tandem.†

Another work that merits mention is entitled "Model Experiments with Variable Camber Wings," by Harris and Bradfield,[9] which was published in 1920.

† Louis Stivers and R. T. Jones of NASA Ames informed me that Chaplygin had done work similar to Lachmann's in 1911 and 1921. See The Selected Works on Wing Theory of Sergei A. Chaplygin. English translation available through Garbell Research Foundation.

The report includes both test data and studies of the significance of the results on airplane performance. Figure 6 shows the biplane cellule that was studied.

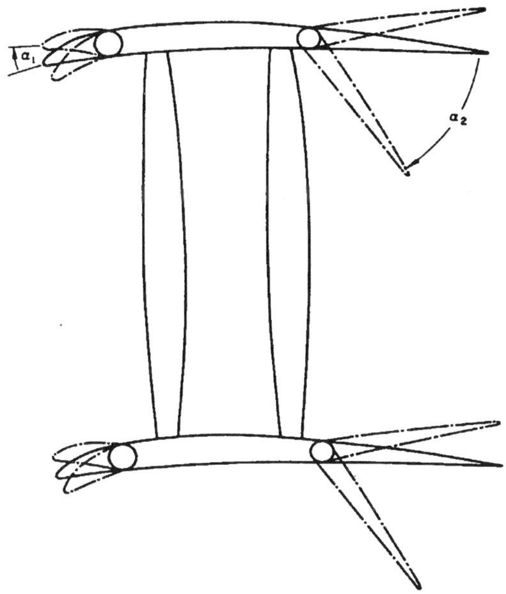

Figure 6
Variable camber Albatross Biplane[9] model, showing approximate range of flap angles tested.

Speaking of biplanes, history seems to be repeating itself. In the early days, because they were biplanes, they had thin wings. These suffered from leading-edge stall, for which nose droop or slats were a cure. Then we advanced to monoplanes and thicker wings, and the leading-edge problem almost disappeared. Later, when jets and higher speed entered the picture, wings again became thin, and the leading-edge problem returned.

Original flap development was motivated by three desired benefits. 1) slower flying speeds, hence shorter takeoff and landing runs; 2) reduction of angle of attack near minimum flying speed; 3) increase of drag, or control of drag, in order to steepen glide angle in approach and reduce floating tendencies. Currently, because of large aircraft noise problems, the emphasis under the third item has changed. We are trying to reduce flap drag in order to reduce thrust requirements and hence the noise.

The split flap was conceived and partly developed in the period 1915-1920 as a means of satisfying these desires. Klemin, Schrenk, and Etienne Royer[2] are important contributors to its development. Orville Wright and I. M. H. Jacobs

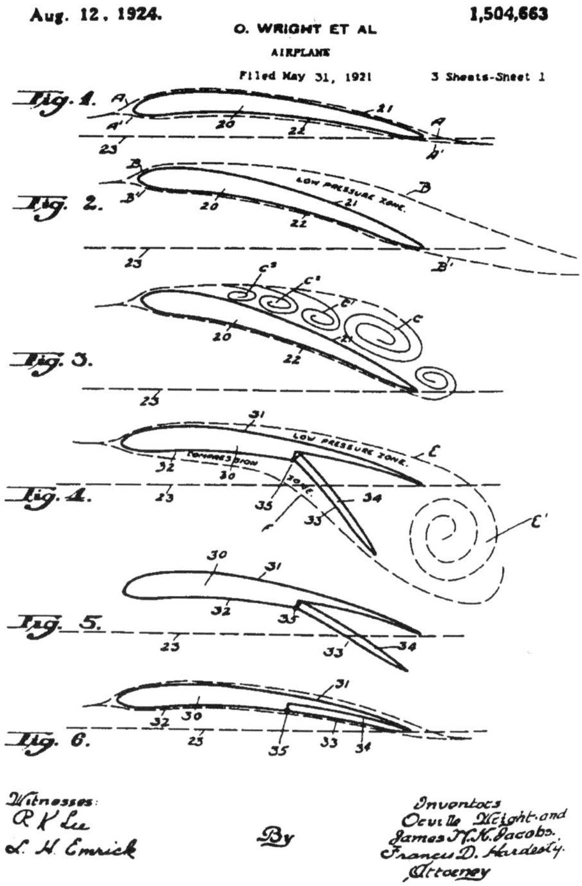

Figure. 7 Sheet I of the U.S. Patent 1,504.663, by Orville Wright and J. M. H. Jacobs, illustrating their concept of a split flap.

obtained a basic patent on the subject in 1924, after they filed it in 1921. Their discussion in the patent gives evidence that they had a good qualitative understandingof the basic flow processes. A reproduction of one of the drawings from the patent is shown in Fig. 7.

We end this historical survey of mechanical high-lift devices by mentioning the Fowler Flap invented by Harlan D. Fowler[10] in 1927. Because it had extension and a slot, it can be considered to be the first modern high-lift mechanical flap. In early NACA wind-tunnel tests,[11] it developed a maximum lift coefficient of 3.17.

Powered lift augmentation, as by suction and blowing, received considerable attention in the 1920's but never quite proved to be efficient enough to justify its use in an actual airplane. Betz[2] and Ackeret[7] in Germany were leaders in this line of development.

For those interested in further research into the early history of high-lift work, we mention several references. First is that by Alston,[12] which was a general lecture to the Royal Aeronautical Society in 1934 entitled, "Wing Flaps and Other Devices as Aids in Landing." It surveys the state of knowledge in 1934, but does not particularly deal with earlier history. Those who are especially interested in the history of high-lift devices and the origin of concepts and applications should see the paper by Weyl,[2] which is a very broad survey—it has 116 references—of the entire subject, with interesting sidelights on the development. As an example of such a sidelight, Weyl notes that in connection with the development of split flaps, a French scientist, Lafay, in 1912 observed "that an unsymmetrical roughening of a cylinder which was exposed to an airflow resulted in an aerodynamic force which was directed towards the smooth side across the flow (lift force)."

The most comprehensive reference examined by the author is a report by A. D. Young,[13] "The Aerodynamic Characteristics of Flaps." It deals with the aerodynamic characteristics—lift, drag, moment, etc., of all types of flaps. Its extensive bibliography covers all aspects, including general reviews, history, theory, and investigations of the various types of devices. In all, 138 references are given. The paper is certainly not obsolete, although it was published in 1953.

A paper that amounts to an updating of Alston's[12] is one by Duddy,[14] which was also presented to the Royal Aeronautical Society. It compares and evaluates various types of flaps, and analyzes their benefits in terms of landing and takeoff performance. Effects of sweep are included.

A useful historical summary of the gradual improvement of gross lift coefficients is shown in Fig. 8, which is due to Cleveland.[15] From about 1935 to 1965—a period of 30 years—we have advanced from coefficients of roughly 2 to roughly 3 on important service-type airplanes. By 1995 will we have advanced to 4?

3. Some Lift Limits

3.1 Limits in Potential Flow

Just as ideal-cycle limits are useful in thermodynamics, so are the theoretical limits of lift useful in aerodynamics. Knowledge of those limits helps give us a perspective as to where we are now and what may be attainable if we are will

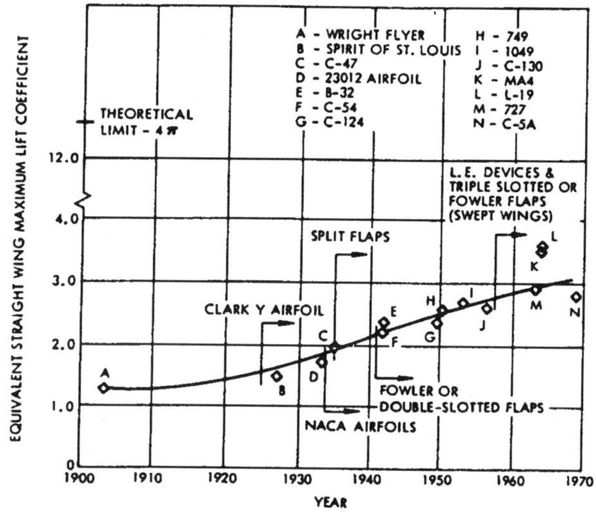

Figure 8 Growth of maximum lift coefficients for mechanical lift systems as a function of time, according to [15].

ing to seek without compromise the maximum possible lift. First consider the limits of c_l in inviscid flow, where separation will not occur. Consider the classical circulatory flow about a circle shown in Fig. 9. Two different levels of circulation are shown; that in Fig. 9b is of the greatest interest, for there the circulation is so strong that front and rear stagnation points coincide. Greater circulation moves the stagnation point off the body. Although that is entirely possible—Flettner Rotor generates such flows—it is not a realistic analog for an airfoil, which in general has two stagnation points, both on its surface. If the reference chord is taken as the diameter of the circle, it is easily shown that flow b of Fig. 9 represents a lift coefficient,

$$c_l = 4\pi \qquad (3.1)$$

where the reference length is the diameter. Now let us consider mean lines. We shall consider only circular arcs, because they are easily obtained from Joukowski's transformation for flow about a circle. In view of the ever present specter of separation, it is an undue refinement to consider either the effects of thickness or the effects of other kinds of mean lines. Three possible circular-arc mean lines, A, B, and C are shown in Fig. 10. A fourth or limiting case is just a straight line. According to Joukowski airfoil theory, for any of those circular-arc mean lines, regardless of degree of camber,

$$L = \pi \rho V_\infty^2 c[\sin(\alpha + \beta)]/\cos \beta \qquad (3.2)$$

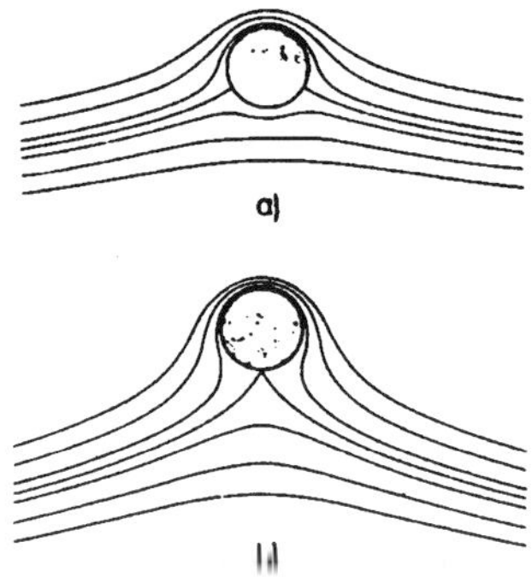

Figure 9 Flow about a circle with two different degrees of circulation. a) moderate circulation, b) circulation so strong that the two stagnation points coincide.

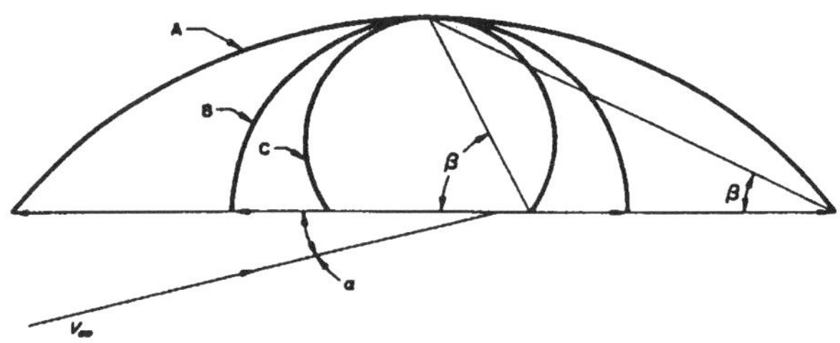

Figure 10 Three circular-arc mean lines, A. B, C, stemming from the Joukowski transformation.

In that equation, c is the length between the ends of the arcs, α the angle of attack, and β is a measure of the camber as shown in Fig. 10. For arcs A and B, c is indeed the chord as conventionally defined. But for arc C, the chord c is not now the longest dimension; the diameter is. In fact, it is evident from the figure that as $\beta \rightarrow 90^0$, $c \rightarrow 0$. Hence, if we define the lift coefficient in terms of the longest dimension, we have [from Eq. (3.2)],

$$c_l = 2\pi\,[\sin(\alpha + \beta)]/\cos\beta \qquad (0 \leq \beta \leq 45^o) \qquad (3.3)$$

$$c_l = 4\pi \sin\beta \sin(\alpha + \beta) \qquad (45^o \leq \beta \leq 90^o) \quad (3.4)$$

Consider Eq. (3.3) with $\beta = 0$. We have a flat-plate airfoil. Its maximum lift coefficient is 2π at $\alpha = 90^o$. A well-known formula applying to symmetrical

Joukowski airfoils that accounts for thickness ratio t is

$$c_l = 2\pi\,(1 + 0.77t)\sin\alpha \tag{3.5}$$

That formula gives a value that is a few percent greater for common values of t.

Next, consider Eqs. (3.3) or (3.4) with $\beta = 45^\circ$. That mean line is a half circle. Maximum lift occurs at $\alpha = 45^\circ$, so that

$$c_{l_{max}} = 4\pi/(2)^{1/2} \qquad (\beta = 45^\circ) \tag{3.6}$$

Finally consider Eq. (3.4). The maximum occurs when $\beta = 90^\circ$ and $\alpha = 0^\circ$, which gives

$$c_{l_{max}} = 4\pi \tag{3.7}$$

The limiting mean line is a complete circle. For airfoils with some thickness, there is a definite stagnation point at the nose and another one at the tail. Hence, that limiting mean line brings the two stagnation points together again, and we effectively recover the flow about a circular cylinder, as in Fig. 9b.

The last is an extreme mean line. The half-circle is not so extreme. In fact, modern multielement flap systems at full flap deflection begin to approximate the half-circle mean line (e.g., Fig. 29). The quantity $c_l = 4\pi/(2)^{1/2}$ in Eq. (3.6) has a value of nearly 9. Of course, the value of c_l depends on the length used as a reference. If arc length were used instead of conventional chord, as above, the values would not appear to be so high. In fact, for the straight line, half circle, and full circle, the theoretical values would be 6.28, 5.65, and 4.0, respectively.

The mean lines that have been considered approximate flows about airfoils having sharp trailing edges. For them, the Kutta condition sets the circulation to such strength that the rear stagnation point is always at the trailing edge. Such a flow might be called a natural flow. To round out our discussion, we shall mention a case in which there is no sharp trailing edge: the ellipse family. Assume the circulation to be controlled in such a way that the rear stagnation point always remains at the rear end of the x-axis, that is, at the same point as in symmetric nonlifting flow (see Fig. 11). Thwaites[16] proposed airfoils that followed that principle. He used area suction at the rear to eliminate separation. When our condition for the circulation is met, the lift coefficient based on the length of the x-axis becomes exactly

$$c_l = 2\pi\,(1 + t)\sin\alpha \tag{3.8}$$

If c_l were based on the length of the y-axis, it would be

$$c_l = [2\pi\ (1+t)/t]\ \sin\alpha \tag{3.9}$$

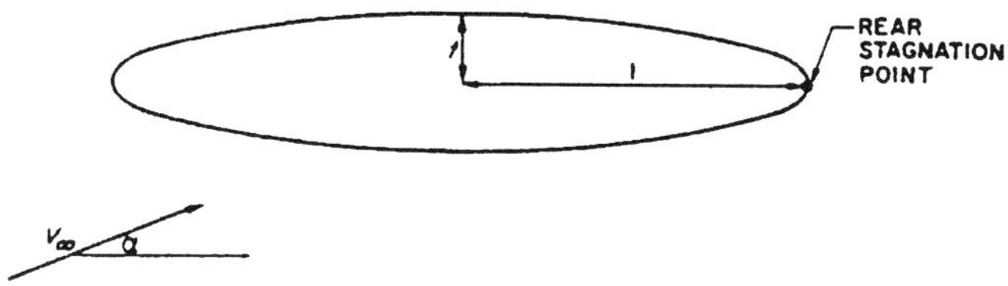

Figure 11 Lifting flow about an ellipse with circulation adjusted to maintain the rear stagnation point at the end of the x-axis.

Observe that Eq. (3.8) is nearly the same as Eq. (3.5). The circular-cylinder case is represented by $t = 1$, and again we recover $c_l = 4\pi$. Equation (3.8) shows that c_l greater than 4π can be exceeded if t is greater than 1, that is, for ellipses that are broadside to the flow. In those cases, we are not using our greatest length as the reference length. We now should use the thickness instead of the length and obtain Eq. (3.9). If Eq. (3.9) is used, $c_{l_{max}}$ falls continuously as t increases. Since both formulas give $c_{l_{max}} = 4\pi$ for a circle, we once more recover 4π as a limit. That is the limiting value for any single-element airfoil. We know that in a uniform stream $L = \rho V_\infty \Gamma$, where Γ is the circulation. Of course, any airfoil can be mapped from a circle. But the mapping does not change the circulation. Hence, under the restrictions we have imposed, the maximum possible lift for any kind of airfoil is 4π.

Since lift is needed near the ground perhaps more than anywhere else, an interesting subsidiary question is the potential flow limit at the ground. The author knows of no simple definitive answer. However, the Douglas Neumann program was applied to a circle that was brought nearer and nearer to the ground, using the image system. The stagnation point was maintained at the bottom as in Fig. 9b. For zero height off the ground the lift coefficient extrapolates to about 4.49, which is also the value found for the classical problem of a round cylinder lying on a streambed. Ground, therefore, greatly reduces the maximum possible lift. We realize full well that ground effect often increases the lift of real wings.

3.2 Limits of $M_\infty^2 C_L$

High values of C_L cannot be maintained indefinitely as speed is increased, for soon surface pressures less than absolute zero would be indicated. Let us look

3.2 Limits of $M^2_\infty C_L$

High values of C_L cannot be maintained indefinitely as speed is increased, for soon surface pressures less than absolute zero would be indicated. Let us look into the problem briefly, and search especially for the limits of lift rather than of lift coefficient. The usual equation for lift is

$$L = \tfrac{1}{2}\,\rho_\infty V^2_\infty C_L S \tag{3.10}$$

An alternate form, one that uses a different expression for dynamic pressure, is

$$L = \tfrac{1}{2}\,\gamma p_\infty M^2_\infty C_L S \tag{3.11}$$

With $\gamma = 1.4$, we can rewrite it as

$$L/p_\infty = 0.7\,(M^2_\infty C_L)S \tag{3.12}$$

Since C_L is known to be a function of M_∞, the product $M^2_\infty C_L$ is the quantity that is of real significance, and so we seek to make statements about its value. Observe that for a given value of $M^2_\infty C_L$ the lift is now proportional to the atmospheric pressure.

A gas cannot be in tension. Hence the limiting suction pressure is a perfect vacuum over the entire upper surface. The limiting pressure on the lower surface is stagnation pressure.

By definition, with $\gamma = 1.4$,

$$C_p = \frac{p - p_\infty}{1/2\rho_\infty V^2_\infty} = \frac{p - p_\infty}{0.7 p_\infty M^2_\infty} \tag{3.13}$$

If the flow is assumed to be isentropic, C_p can be written in the form

$$C_p = \frac{1}{0.7M^2_\infty}\left[\left(\frac{1 + 0.2M^2_\infty}{1 + 0.2M^2}\right)^{3.5} - 1\right] \tag{3.14}$$

Before proceeding to the determination of lift limits, it is interesting to pause and consider the limits of C_p values. In high-lift testing, maximum lift is often found to occur when the highest local velocities reach a Mach number of 1.0. The

C_p corresponding to that condition will be called C_p^* . If a perfect vacuum were reached, we would obtain, according to Eq. (3.13), a value of $M_\infty^2 C_p$ equal to – -1.43. Mayer[17] looked for the highest possible negative C_p values in experiment by examining hundreds of NACA test data points from all sorts of tests. He found the remarkable empirical result that the highest experimentally measured

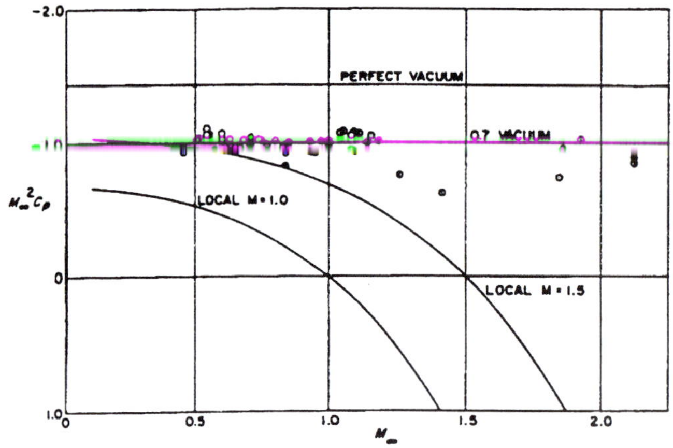

Figure 12 $M_\infty^2 C_p$ values as a funcion of Mach number for four conditions. 0.7 vacuum corresoponds to $M_\infty^2 C_p$ = - 1.

values correspond with reasonable accuracy to $M^2 C_p$ = – 1. That value corresponds to 0.7 of a vacuum [see Eq. (3.13)]. Figure 12 shows the test data he used to support the finding. Since much of our interest is in high-lift testing, which is done at low Mach numbers, we convert some of the results from Fig. 12 into C_p form and present them in Table 1.

Table 1 Limiting values of C_p at low Mach numbers

M_∞	0.10	0.15	0.20	0.30	0.40	0.50
C^*_p (M = 1)	-67	-29.3	-16.3	-7	-3.7	-2.1
C_p (perfect vacuum)	-143	-63.6	-35.8	-15.9	-8.9	-5.7
C_p (0.7 vacuum)	-100	-44.5	-25	-11.1	-6.3	-4.0

At low Mach numbers, quite high values of C_p are obtainable before compressibility becomes very important, but at higher Mach numbers the limiting C_p values are not very high. Note, however, that Mayer's data in Fig. 12 do not cover the lower Mach numbers; hence their validity there is not truly established.

The author knows of no theoretical explanation for Mayer's 0.7 vacuum correlation. It does not correspond to a constant local Mach number. By the use of Eq. (3.14), it is easily deduced that $M_\infty^2 \, C_p = -1$ corresponds to $M_{local} = 1.43$ at $M_\infty = 0$ and 1.55 at $M_\infty = 0.5$.

Now consider the airfoil problem. If we hold M constant at one value on the upper surface and constant at another value on the lower surface, C_p represents the C_L for each surface. The total C_L is the difference. Hence from Eq. (3.14),

$$M_\infty^2 \, C_L = \frac{1}{0.7} \left[\left(\frac{1 + 0.2M_\infty^2}{1 + 0.2M_l^2} \right)^{3.5} - \left(\frac{1 + 0.2M_\infty^2}{1 + 0.2M_\mu^2} \right)^{3.5} \right] \tag{3.15}$$

where l and μ denote lower and upper surfaces. A perfect vacuum on the upper surface corresponds to $p_\mu = 0$ or $M_u = \infty$. In that case, the second term in Eq. (3.15) is zero. With respect to atmospheric pressure, the load carried by the upper surface when there is a perfect vacuum is [see Eq. (3.13)].

$$M_\infty^2 \, c_l = 1/0.7 = 1.43, \text{ a constant} \tag{3.16}$$

If the entire flow on the lower surface is a stagnation flow, $M_l = 0$, and hence with a vacuum on the upper surface we have

$$M_\infty^2 \, C_L = (1+0.2M_\infty^2)^{3.5}/0.7 \tag{3.17}$$

Equation (3.17) represents the absolute lifting limit. The quantity increases with Mach number because lower-surface pressures can increase with Mach number. Equation (3.17) is optimistic, because isentropic compression is assumed, whereas in reality there surely will be a shock when M » 1. Table 2 shows values of $M_\infty^2 C_L$ for several flow conditions, including Mayer's $M_\infty^2 C_L = 1$ limit. The maximum value possible in subsonic flight is 2.71, and the contributions from each surface are about equal. If Mayer's value $M_\infty^2 C_L = -1$ is accepted, the limit decreases to 2.28. At $M_\infty = 0.5$, the absolute maximum value is only 1.70 and Mayer's value is 1.27. Those are the limiting values, regardless of the kind of high-lift devices that are used. At $M_\infty = 1.0$ and below, the assumption of isentropic recompression on the lower surface is very good, as is well known. As Mach numbers become large, pressures on the lower side become great, and large values of $M_\infty^2 C_L$ develop—values that exceed 11 at $M_\infty = 2.0$. Mayer considers the case where under-surface pressures in supersonic flight correspond to a normal shock. He gives considerable information on maximum-lift values in supersonic flight, including correlations of theory and experiment. But since our interest is principally in subsonic flight, we shall not discuss his work further.

3.3 Demonstrated Lifting Limits

It is interesting that the question of aerodynamic lifting limits in subsonic flight was posed to the author in 1946 by E. H. Heinemann. The problem was fighter maneuvering—what lift could a wing really develop? All the available flight and wind-tunnel data were examined. The examination included the large supply of German World War II data, which covered tests of a great variety of airplane and missile-type wings. The results were reported in Ref. 18. The study, which was strictly one of observation, concluded that the maximum possible lift was about 1/2 atm. That number corresponds to a value for $M_\infty^2 C_L$ of about 0.7.

Since that time, the design of swept wings has advanced considerably. Now, the highest value of $M_\infty^2 C_L$ seen by the author is 1.20 on a swept wing using an aft-loaded airfoil section with no auxiliary lift devices. The value corresponds to $C_L = 1.26$ at $M_\infty = 0.975$. It was not really a maximum-lift value but, rather, a stopping point in a wind-tunnel test that was determined by strength of the sting support. Hence the value can surely be exceeded. It should be pointed out that this is not a buffet limit, which is usually considerably lower. If one went all out to maximize $M_\infty^2 C_L$, without regard to low-lift performance, much better could surely be done. Leading-edge and trailing-edge devices undoubtedly would raise the limit. Variable-camber wings are currently receiving attention as another possible means of raising the limit.

Some further data related to limiting lift conditions are shown in Figure 13 which is taken from Ref. 19. In tests of a considerable variety of two-dimensional models, it was observed that the lowest C_p on the model, at $c_{l_{max}}$ was nearly a constant equal to C_p^* . There is considerable scatter in the data of Fig. 13 because the tests were not planned to investigate C_p at $c_{l_{max}}$, but there is considerable of a critical condition. If compressible C_p values can be calculated for an airfoil, an easy way of estimating $c_{l_{max}}$ is suggested by the figure—just find the C_l at which C_p becomes sonic. All the tests were on airfoil sections where rear separation was not critical.

Table 2 Maximum-lift limits assuming isentropic compression, and uniform chordwise loading

M_∞^2	M_{upper}	$M_\infty^2 \; C_{L_{upper}}$	M_{lower}	$M_\infty^2 \; = \; C_{L_{lower}}$	$M_\infty^2 \; C_{L_{total}}$
0.5	∞	1.43	0	0.27	1.70
0.5	1.55	1.00	0	0.27	1.27
0.5	1.5	0.97	0	0.27	1.24
1.0	∞	1.43	0	1.28	2.71
1.0	1.86	1.00	0	1.28	2.28
1.0	1.5	0.69	0	1.28	1.97
2.0	∞	1.43	0	9.75	11.18
2.0	4.97	1.00	0	9.75	10.75

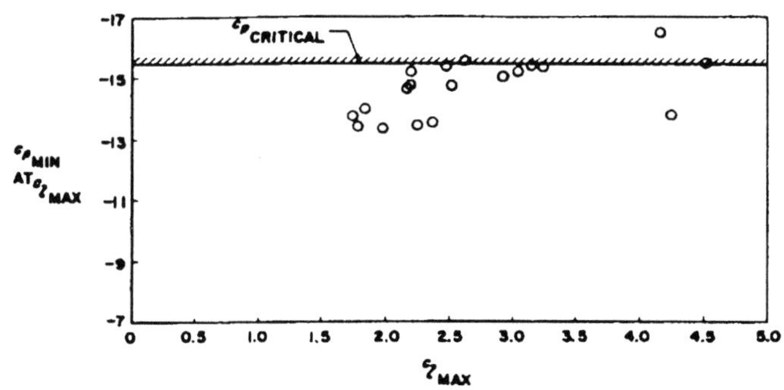

Figure 13 Experimentally measured leading edge minimum pressure coefficient at $c_{l_{max}}$ for a variety of two-dimensional high-lift configurations. Test Mach number = 0.20.

4. The Load-Carrying Capacity of Natural Boundary Layers

4.1 Introduction

This and the next section constitute the body of the paper. In combination, they expound the basic aerodynamics of mechanical high-lift systems. Until the 1960's, our analytical capability was insufficient to make quantitative calculations. Now, since the tools are here, it is time to put the whole story together, as best we know it. That is not to say that nearly all problems have been solved, but to a certain extent the remaining problems amount to just irritating details.

We go into considerable detail, and often rather elementary exposition, in the interests of making this part of the paper complete. The approach is motivated by the many misconceptions the author has encountered. For instance, with respect to slotted flaps, it has been suggested that if one just knew how to shape the airfoil, higher lift could be obtained with a one-piece airfoil, because the leakage from bottom to top allowed by the slots amounts to a kind of short circuit. Properly designed multielement airfoils are better, as will be shown. Another common statement is that a slot provides a kind of blowing boundary-layer control by the jet of "fresh" air that flows through it. How can it do that? It has no more total head than the ambient stream. Admittedly, that jet of "fresh" air has more energy than the boundary layer, but it only has freestream total head. At best, it may be a question of semantics. Later we shall see that the principal effect of a slot or slat is to reduce negative pressures, rather than to blow.

Sections 4 and 5 are a revision and, we hope, an improvement of Ref. 20, a paper given at an AGARD meeting. There are many papers and reports on the subject of high-lift devices, but their emphasis usually tends to be "how to." Ours is "why." One useful document is Ref. 21, which contains a large collection of lectures on the subject of high-lift devices. Another is Thain's,[22] which is an extensive review of the experimentally observed effects of Reynolds number on high lift. It too is more concerned with overall results than with explanation of the fundamental flow processes. Let it be noted that our discussion in these sections is confined to two- dimensional flow.

In view of Bernoulli's law, if a surface is to lift, the velocity over the upper surface must be speeded up. But we know that for any trailing edge, even the cusped type, the flow decelerates to below freestream values. Hence, to get more lift, we need higher velocities over the airfoil, but that in turn means greater deceleration towards the rear. The process of deceleration is critical, for if it is too severe, separation develops. The science of developing high lift, therefore, has two components: 1) analysis of the boundary layer, prediction of separation, and determination of the kinds of flows that are most favorable with respect to separation; and 2) analysis of the inviscid flow about a given shape with the purpose of finding shapes that put the least stress on the boundary layer. The two parts amount to a kind of applied-load and allowable-load problem. An analysis cannot be valid unless its elements are sound, so let us first look at the problem of predicting the onset of separation.

4.2 Accuracy of Predicting Separation Points

Because this question is so vital to many aerodynamic design problems, the accuracy of four leading methods was studied recently in Ref. 23, for the case of turbulent flows, which is the one of interest. The flows considered involved simple rear separation. Laminar bubble or laminar leading- edge separations were not considered. The main reasons are that proper design can eliminate them and that at high Reynolds numbers transition has occurred by the time adverse gradients appear. Fortunately, the problem we consider here is technically the more important, because the problem of separation after a laminar bubble and reattachment cannot be analyzed nearly as well. That is a problem for the future. There is no difficulty with simple laminar separation. It can be calculated to three- or four-place accuracy if the pressure input data have sufficient accuracy.

Four leading methods were examined: those of 1) Goldschmied, 2) Stratford, 3) Head, and 4) Cebeci-Smith. One case representative of the fairly large number examined in Ref. 23 is shown in Fig. 14. The separation point was carefully measured. The predicted separation points are marked on the pressure-distribution curves, which were supplied by the experiments. Goldschmied's method was erratic. The other three were in reasonable agreement, both in what is shown here and in the complete study[23], Stratford's method tended to predict separation slightly early. The Cebeci-Smith method appeared best, with the Head method a strong second.

Hence, the Cebeci-Smith method has been chosen as the basic method,

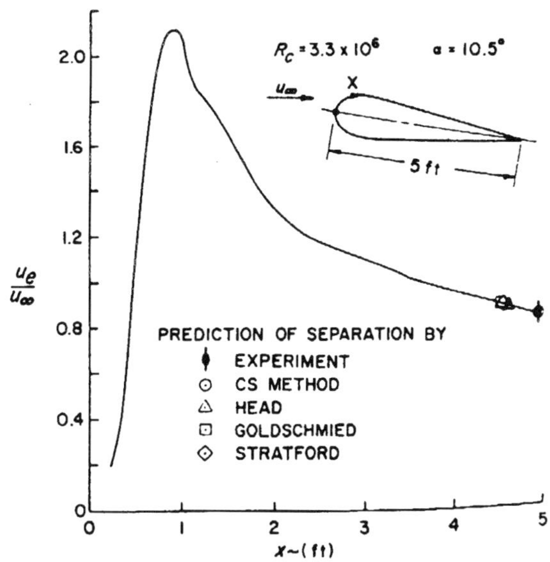

Figure 14 Comparison with experiment of predicted separation points for Newman's airfoil.

although much use is made of Stratford's method and ideas in this paper, mainly
because of their great convenience. Figure 15 summarizes the prediction accuracy of the Cebeci-Smith method. The data consist of the results from Ref. 23 together with other, unpublished studies. One should not conclude from Fig. 15 that the general accuracy of turbulent-boundary-layer calculation is as good as it appears to be. Near separation, the accuracy of many of the boundary-layer quantities, such as the momentum thickness θ and the velocity profile, become poor. Nevertheless, accuracy of predicting the location of separation remains good. The Cebeci-Smith method predicts separation to be at the point where c_f, the local skin-friction coefficient, equals zero.

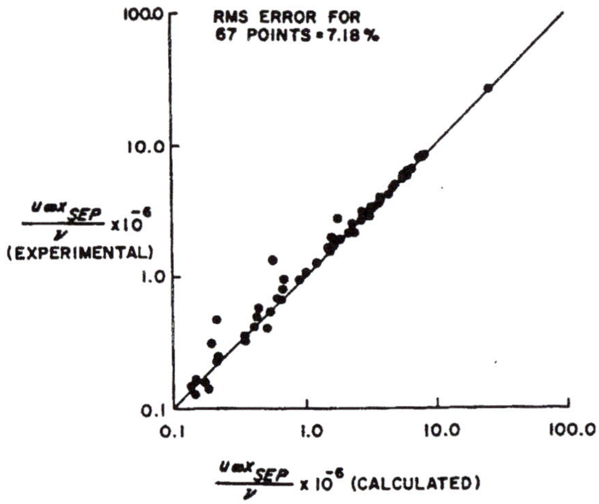

Figure 15 Accuracy of predicting turbulent separation points by the Cebeci-Smith method.

In view of those results, the method of predicting separation appears to be accurate enough, in an engineering sense, to be used in determining relations, limits, and the like. It should be noted that the Cebeci-Smith method can handle with great accuracy the effects of Reynolds number or Mach number up to 5 or more. Also, it is the only one of the four methods that can analyze the case of axisymmetric boundary-layer flow.

4.3 Canonical Pressure Distributions

In almost all design work, pressure distributions are presented in terms of $C_p = (p - p_\infty)/(1/2\rho u_\infty^2)$. In that kind of presentation, high negative values of C_p invariably look bad, and one is unable to tell by inspection much about the margin of safety of the boundary layer against separation. Yet we know from basic scaling considerations that if two pressure distributions can be made congruent by proper scaling in the x- and C_p-directions, then the two flows are identical except for the Reynolds number effect, which is weak. Then, if separation occurs, it will be at the same scaled point for both flows. A particular 2-in. airfoil model at 100 mph will have very high values of velocity gradients, but a similar 200-in. model at 1 mph will have extremely low values. Yet the flows are exactly similar because their Reynolds numbers are the same. It is the dimensionless shape that counts. Hence it is particularly useful to scale out the magnitude of the velocity and also to scale out the chord. Where separation is important, the best scaling factor is the velocity just before deceleration begins. Because all pressure distributions are put in a standard form, it is natural to call them canonical pressure distributions and use a new variable $\bar{C}_p$. A typical one is illustrated in Fig. 16, which shows the idea and basic relations. The exact details of the normalization may well depend on the problem and the nature of the pressure distribution. The canonical pressure distribution, together with a Reynolds number, completely describes the flow. A meaningful Reynolds number is R_θ at the beginning of pressure rise. The left-hand part in Fig. 16 might represent the nose of an airfoil. Distance x is measured along the surface, but the origin of x is a matter of convenience. Often it is convenient to locate it at the beginning of pressure rise, as in the figure. Separation may occur at some point as noted. Because of the very simple relation between pressure coefficient and velocity ratio, the term pressure distribution is used indiscriminately for plots of either pressure coefficient or velocity ratio, provided that the flow is at low speed.

In the canonical system $\bar{C}_p = 0$ represents the start of pressure rise and $\bar{C}_p = +1$ the maximum possible, that is, $u_e = 0$. Normally there are no negative values

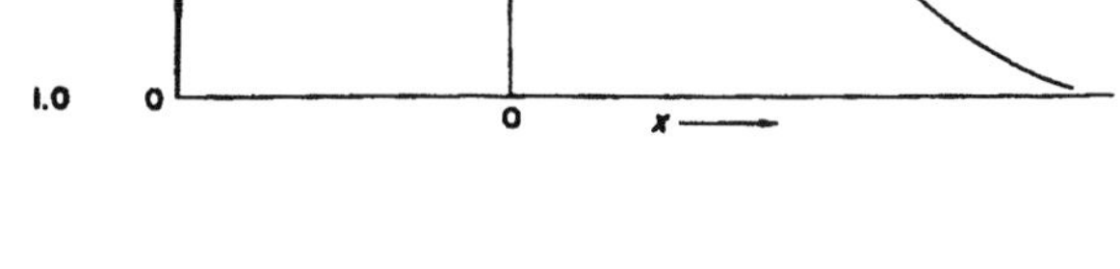

Figure 16 A canonical pressure distribution, $\bar{C}_p$ (x).

of $\bar{C}_p$. Furthermore, if two pressure distributions can be made congruent by proper scaling, a flow having a deceleration of $(u_e/u_\infty)^2$ from 20 to 10 is no more and no less likely to separate than one decelerating from 1.5 to 0.75 or even from 0.10 to 0.05. The canonical plot is the one that is meaningful in separation analysis. With magnitude effects scaled out, much more can be told by a simple inspection than by a conventional plot. We dwell on the canonical pressure distribution at some length, because most working aerodynamicists do not realize its value. Figures 17 and 18 show conventional and canonical pressure plots for a typical high-lift airfoil. The basic character of the four pressure distributions is much more apparent in Fig. 18. Figure 17 includes a canonical scale at the right, for the complete ensemble.

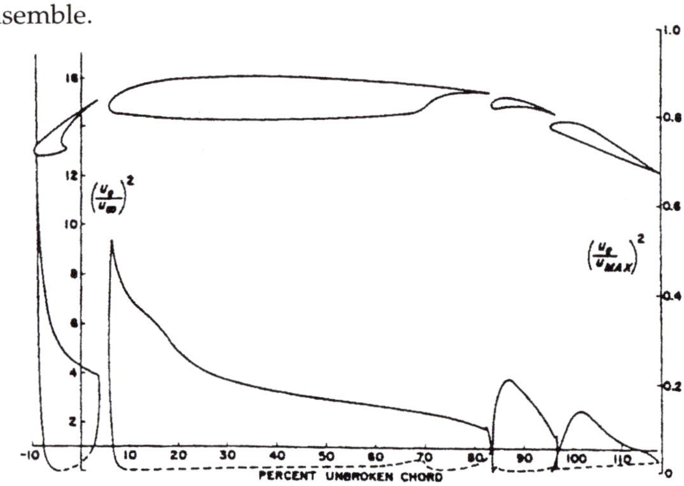

Figure 17 A conventional theoretical C_p plot of a four-element airfoil. The right-hand scale is the canonical form, referred to peak velocity at the nose of the flap, α = 15°, δ_f = 25°, δ_s = 25°, c_l = 4.08.

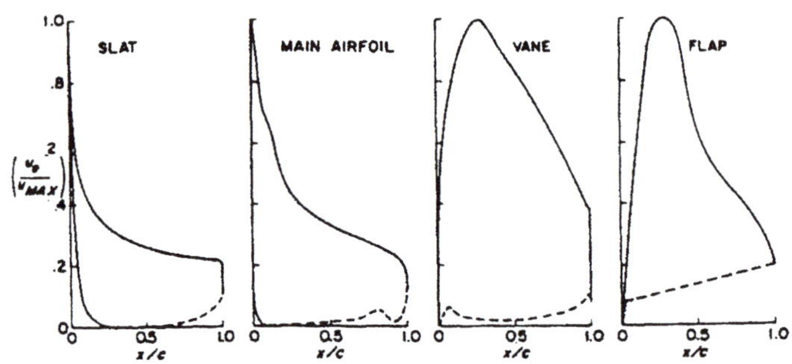

Figure 18. Airfoil C_p of Fig. 17 plotted in canonical form.

4.4 Relation between the Canonical Pressure Distribution and the Conventional

The relations between canonical and conventional pressure distributions are very simple, but because there has been confusion on the subject, they will be discussed at some length here. The canonical pressure distribution is easily related to the conventional form, where the reference velocity and pressure are u_∞ and p_∞. Because the square of the velocity is never negative, it is simpler to deal with velocity than with pressure coefficient. The fundamental relation is quite simple. It is

$$\left(\frac{u_e}{u_\infty}\right)^2 \left(\frac{u_e}{u_\infty}\right)^2 = \left(\frac{u_e}{u_o}\right)^2 \bullet \left(\frac{u_o}{u_\infty}\right)^2 \tag{4.1}$$

The last factor is just a constant. In many design problems, it is desirable to start with a canonical pressure distribution, then (u_o/u_∞) is not necessarily known and must be found. Construction of a canonical pressure distribution from a conventional one by means of Eq. (4.1) is trivial, because u_o/u_∞ is known. The method of converting velocities into C_p or $\bar{C}_p$ form is obvious from the given relations.

A problem of some importance is that of applying a canonical pressure distribution to the design of an airfoil by inverse methods. Somewhere along the pressure distribution the flow may separate. Then in the adaptation to the airfoil, the separation point should not be forward of the trailing edge. From Eq. (4.1) we can get the following equation:

$$\left[\frac{u_e}{u_\infty}\right]^2 (x) = \left[\frac{u_e}{u_o}\right]^2_{\text{sep}} \left[\frac{u_0}{u_\infty}\right]^2 \tag{4.2}$$

Solving for $(u_0/u_\infty)^2$, substituting into Eq. (4.1), we get

$$\left[\frac{u_e}{u_\infty}\right]^2 (x) = \frac{(u_e/u_\infty)^2_{sep}}{(u_e/u_0)^2_{sep}} = \left[\frac{u_e}{u_0}\right]^2 (x) \tag{4.3}$$

From general airfoil theory, we know that $(u_e/u_\infty)^2$ at the trailing edge is about 0.8, corresponding to $C_p = 0.2$. Then, for instance, if the canonical pressure distribution separates at $(u_e/u_0)^2 = 0.4$, the factor is $0.8 \div 0.4 = 2.0$. Of course Eq. (4.3) is valid, no matter where separation occurs, but an airfoil is not likely to be deliberately designed to have separation. If we assume separation is always to be at the trailing edge for any airfoil designed for maximum lift, we can specialize Eq. (4.3) to the form

$$\left[\frac{u_e}{u_\infty}\right]^2 (x) = \frac{(u_e/u_\infty)^2_{T.E.}}{(u_e/u_0)^2_{sep}} \left[\frac{u_e}{u_0}\right]^2 (x) \tag{4.4}$$

Observe that if by some artifice we could greatly increase $(u_e/u_\infty)^2_{T.E.}$, we could greatly increase $(u_e/u_\infty)^2(x)$ and hence the lift, all for the very same canonical pressure distribution. More will be said about that later. Equation (4.4) is easily converted to C_p form. Writing

$$C_p = 1 - (u_e/u_\infty)^2,\ C_{pT.E.} = 1 - (u_e/u_\infty)^2_{T.E.}$$

and

$$\bar{C}_p = 1 - (u_e/u_0)^2 \tag{4.5}$$

we have

$$C_p(x) = 1 - \frac{1 - C_{pT.E.}}{1 - \bar{C}_{psep}} \left[1 - \bar{C}_p(x)\right] \tag{4.6}$$

4.5 Location of Separation in Two Families of Canonical Distributions

To exhibit the limits on the ability of a boundary layer to flow into regions of higher pressure, we present two plots, covering families of flows at two different Reynolds numbers, Figs. 19 and 20. The flows consist of a length of constant-velocity flow followed by a pressure rise described by the equation

$$\bar{C}_p = x_m \tag{4.7}$$

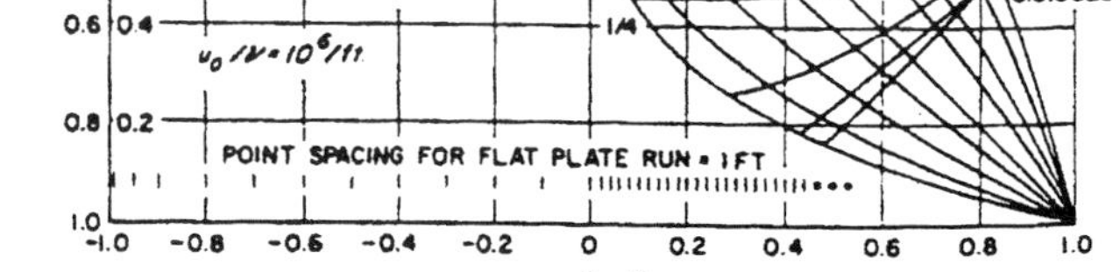

Figure 19 Separation loci for a family of canonical pressure distributions. Point spacing used in the boundary-layer calculations for the 1-ft rooftop run is noted. Values in parentheses under origins of flow are values of R_q at $x = 0$.

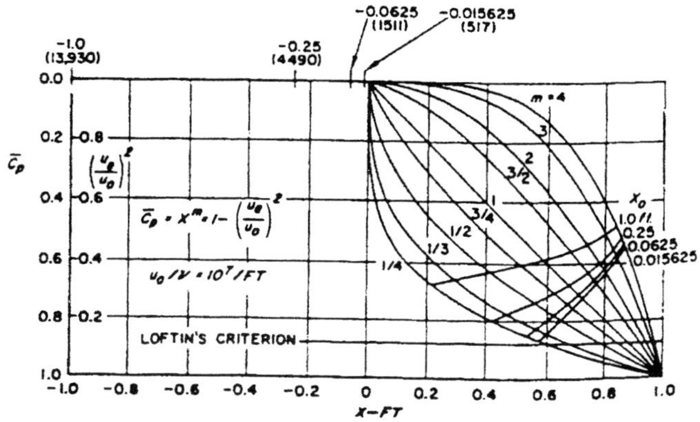

Figure 20 Separation loci for a family of canonical pressure distributions. Loftin's criterion $\bar{C}_p$= 0.88 is noted.

The unit Reynolds number is 10^6 per ft in Fig. 19 and 10^7 per ft in Fig. 20. Pressure rise is set to start at $x = 0$, but forward of that point are various lengths of flat-plate flow. It seemed more convenient to construct the plots in terms of feet and u_∞/ν rather than in terms of Reynolds number, although conversion into Reynolds number is easy. The total region of pressure rise is seen to be 1 ft. Four lengths of flat-plate run were studied: 1/64 ft, 1/16 ft, 1/4 ft, and 1 ft. The initial flows then developed boundary layers of various thickness at the beginning of pressure rise, as indicated by values of R_θ in the figures. The flat-plate flow is assumed to be entirely turbulent. If it were mixed laminar and turbulent, values of R_θ at $x = 0$ would be less and different.

Calculations of the flat-plate and $\bar{C}_p = x^m$ parts of the flow were then made by the Cebeci-Smith method until separation was reached. Lines cutting across the $\bar{C}_p = x^m$ curves mark separation points for the four lengths of flat-plate runs. The straight-line or concave pressure rises permit the greatest recovery before separation occurs. Also, the curves show that the amount of recovery is sensitive to the length of flat-plate run before the beginning of recovery. The separation loci, while indeed functions of the length of flat-plate run, are more fundamentally functions of the boundary-layer thickness at the beginning of pressure rise. Hence, the separation loci could just as well be identified by the noted values of R_θ at the beginning of pressure rise. Then Figs. 19 and 20 become applicable for any com-bination of rising pressure and transition. For any arbitrary forward flow, one would then calculate R_θ at $x = 0$ and interpolate between the loci as a function of R_θ. In brief, we are saying that any flow that develops one of the set of values of R_θ and that has the same aft flow will have the same separation point as that of a forward flow developed by an equivalent run of flat plate. The equivalence is not rigorous, because a separation point is a function not only of initial Re but also of the shape of the profile as described by H. But R_θ is the dominant parameter, and, considering the accuracy of turbulent boundary-layer calculations, it is the only one that need be considered here. With u_o/v specified, θ converts to R_θ as the interpolation parameter.

These comments point out the fact that extensive laminar low in the forward part helps to delay separation considerably, because it reduces R_θ. Furthermore, boundary-layer suction forward of the pressure rise would delay separation. Another fact that is surprising is that if the pressure rise is not too great, it may be extremely steep; in fact, $d\bar{C}_p/dx$ may be nearly infinite. (Owing to the use of finite steps, the slope of $\bar{C}_p$ is never quite infinite in the Cebeci-Smith method. If it were, a constant in the partial differential equation would be infinite and the method would fail.) In the next section, we see that the infinite rate of pressure rise is confirmed by Stratford's equations, which he derived by an analytic approach.

Loftin and von Doenhoff[24] studied a large number of thin airfoils and arrived at a separation criterion that in our terms is $\bar{C}_p = 0.88$. It is plotted in Fig. 20. A thin airfoil at angle of attack has an effective forward flow about the same as that of the $x = 1/64$-ft distance, and the pressure-rise region is approximated by the $m = 1/3$ or $m = 1/4$ curve. Hence, we see that Loftin's criterion is rather closely predicted by the charts. But we also see that it is far from universal.

The set of charts provides an "eyeball" method for estimating separation points. We shall illustrate by considering Fig. 14. Try to relate that flow to one of the canonical plots. If the pressure distribution of Fig. 14 is replotted in u_e^2/u_∞^2, or canonical form, the rear flow looks approximately like the $x^{1/3}$ flow and the front flow is effectively quite short. Transition was measured at $x = 1.169$ ft. Hence the logical equivalent flat-plate length is 0.015625 ft. Which chart should be used? Probably Fig. 20. To use the figures, convert the chord of the airfoil to 1 ft. Because it is so small, there is no need to be concerned about the 0.015625 forward part. In order to hold Reynolds number at 3.3 x 10^6 with the reduced chord, we need to increase u_∞ / v to 3.3 x 10^6/ft. But the charts are in terms of u_o / v. Now from Fig. 14, $u_o / u_\infty = 2.1$. Hence, $u_\infty / v = 3.3 \times 10^6$ corresponds to $u_o / v = 7 \times 10^6$. Then Fig. 20 seems to be the better of the two to use. Of course, interpolation between the two is possible. From Fig. 20 for $m = 1/3$, we find that separation occurs at $(u_e/u_o)^2 = 0.17$ or $u_e/u_o = 0.41$. In Fig. 14 $(u_e/u_\infty)_{max} = 2.1$. Hence separation should occur at about 0.41

x 2.1= 0.87, a ratio that is satisfactorily close to the measured separation point. More care in the analysis probably would increase the accuracy. For instance, a careful calculation of R_θ at $x = 0$ by Truckenbrodt's method would better than a guess. Also, the exponent in the pressure-rise region could be determined more accurately by the use of log graph paper.

The canonical plots contain much useful information. For instance, if load is being carried by an airfoil in cruise, the pressure rise at the trailing edge is not great. If it corresponds to $\bar{C}_p = 0.4$, any m-curve will sustain the pressure rise, and therefore suitable airfoils with all kinds of pressure distributions can be made. But if one is striving for all the lift he can get, the pressure-rise curve for $m = 1/4$ or $1/3$ is the best, because that gives the greatest ratio and the highest mean value of $\bar{C}_p$ along the upper surface.

Reference 20 gives two charts for a different family of pressure distributions; they are arcs of circles. Results are not significantly different from those of the $\bar{C}_p = x^m$ family. Furthermore, Ref. 20 contains an additional pair of charts for the case where $\bar{C}_p = x^m$ at $M_0 = 1$. The separation loci are considerably different, but more of the difference appears to be eliminated if we consider the flow

$$1 - (u_e/u_o)^2 = x^m \tag{4.8}$$

The expression for $\bar{C}_p$ when the flow is compressible is

$$\bar{C}_p = \frac{2}{\gamma M_o^2}\left\{\left[1 + \frac{\gamma - 1}{2} M_o^2 \left(1 - \frac{u_e^2}{u_o^2}\right)\right]^{\gamma/(\gamma-1)} - 1\right\} \tag{4.9}$$

If $u_e/u_o = 0$, $\bar{C}_p = 1.28$ at $M_o = 1$; and if $\bar{C}_p = 1.0$, $(u_e/u_o)^2 = 0.18$. Hence, at $\bar{C}_p = 1.0$, the flow is still far from stagnation. Therefore, the better canonical compressible flow to consider is the flow in terms of $(u_e/u_o)^2$. Some points calculated thus for flow at $M_o = 1.0$ are shown in Fig. 8 of Ref. 20. The answer to the question of whether compressibility aggravates separation depends on the reference. Certainly, compressibility aggravates pressure gradients on a given body, which implies earlier separation. When considered in $\bar{C}_p$ form, compressibility is favorable. Hence the most fundamental approach appears to be to consider the problem in terms of $(u_e/u_o)^2$ which is a measure of the kinetic energy of the flow that remains. In that form, at least to $M_0 = 1.0$, the effects of compressibility are minor when considered from the basic canonical standpoint.

In closing this section, we remark that because of the analytic nature of our canonical pressure distributions $\bar{C}_p = x^m$, it is very easy to apply Stratford's criterion. That has been done. If Stratford's predictions had been added to the plots of Figs. 19 and 20, the differences in separation loci would appear considerable. Still they are so small that use of the charts and the findings is not negated. Addition of Stratford-type loci would only cause confusion. Therefore, we show only one type of calculation, the Cebeci-Smith, which in general is believed to be the most accurate. Furthermore, it should be noted that Figs. 19 and 20 differ from their earlier forms in Ref. 20. The main reason is that here we used the analytic nature of the $\bar{C}_p = x^m$ flows to compute certain necessary derivatives.. In the earlier work, finite-difference formulas were used. Disagreements such as those just indicate the state of the art of turbulent-flow calculation.

4.6 Limiting Canonical Distributions

The $\bar{C}_p = x^m$ families are very useful for almost all practical flow problems, but of course the shapes of the $\bar{C}_p\,(x)$ curves are arbitrary; the shapes are selected as a matter of analytical convenience. Just as in many other problems, there is one shape that is "best," Stratford's solution. We do not mean to imply that the solution is exact, but as is indicated by Fig. 14, it has acceptable accuracy. Stratford has derived a formula, Ref. 25, for predicting the point of separation in an arbitrary decelerating flow:

$$\frac{\bar{C}_p\,[x(d\bar{C}_p\,/dx)]^{1/2}}{(10^{-6}R)^{1/10}} = S \tag{4.10}$$

where if $d^2p/dx^2 \geq 0$, then S = 0.39; or if $d^2p/dx^2 \leq 0$, then S=0.35. Also, $\bar{C}_p \leq 4/7$. The flows examined consist first of a flat-plate flow, just as with $\bar{C}_p = x^m$ flows. Hence, x is distance measured from the leading edge of the plate, and $R = u_o x/\upsilon$. If the flows begin the pressure rise at a point x_o such that $\bar{C}_p = (x-x_0)^m$, left-hand side of Eq. (4.10) starts from a zero value, provided that $m > 1/3$. The left-hand side then grows. When it reaches the limiting value of S, separation is said to occur. The study of Ref. 23 showed the constants to be slightly different, but here, for our purposes, we accept Stratford's values. See Ref. 25 for further details.

If S is held at its limiting value of 0.39 for $d^2p/dx^2 \geq 0$, Eq. (4.10) amounts to an ordinary differential equation for $\bar{C}_p\,(x)$. It is evident from Eq. (4.10) that the equation describes a flow that is ready everywhere to separate. Stratford presents the following solutions:

$$\bar{C}_p = 0.645\,\{0.435\,R_o^{1/5}\,[(x/x_o)^{1/5} - 1]\}^{2/n} \tag{4.11a}$$

$$\text{for}\quad \bar{C}_p \le (n-2)/(n+1)$$

and

$$\bar{C}_p = 1 - \frac{a}{[(x/x_o) + b]^{1/2}} \quad \text{for } \bar{C}_p \ge \frac{n-2}{n+1} \tag{4.11b}$$

In that two-part solution, x_o is the start of pressure rise, $R_o = u_o x_o / \upsilon$, x is the distance measured from the very start of the flow, which begins as flat-plate, turbulent flow. The number n is a constant that Stratford finds to be about 6. The quantities a and b are arbitrary constants used in matching values and slopes in the two equations at the joining point, $\bar{C}_p = (n\text{-}2)/(n\text{+}1)$. Of course, Eq. (4.11a) describes the beginning of the flow, and Eq. (4.11 b) the final part. The flow is an equilibrium flow that always has the same margin, if any, against separation. Two families of such flows have been computed; they are shown in Fig. 21. They correspond to the same set of conditions that were used in the

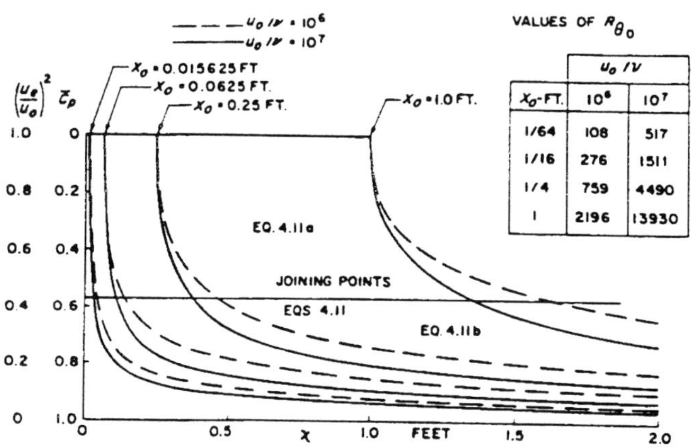

Fig. 21 Stratford limiting flows at two values of unit Reynolds number.

$\bar{C}_p = x^m$ flows. As has already been mentioned, they represent true limiting flows—the slightest increase in adverse gradient anywhere should cause separation. The curves assume the length of flat-plate runs indicated, but just as for the $\bar{C}_p = x^m$ families, what is of more fundamental significance is the boundary-layer thickness at the beginning of pressure rise. Hence, a table of initial values of R_θ is included. With the aid of those R_θ values, the curves could be applied to arbitrary forward pressure distributions having mixed laminar/turbulent flow. Within the accuracy of the theory, which tests have proved to be correct and somewhat conservative, those curves show the fastest possible pressure rise that can be obtained from a natural boundary layer. Also, when compared with Cebeci-Smith predictions, the theory appears conservative. Figure 21, together with Eq. (4.11), exhibits the following features:

1) The initial slope $d\bar{C}_p/dx$ is infinite, so that small pressure rises can be made in distances from very short to zero.
2) It is easy to show that $\bar{C}_p \sim x^{1/3}$ in the early stages.
3) The dominant variable is x/x_0 [see Eq. (4.11)]. Hence, when x_0 is small (i.e., the boundary layer is thin), pressure recoveries may be very rapid. When the initial run is long and the boundary layer is thick, the allowable average pressure gradient is much less. Or conversely, thick boundary layers are much more likely to separate than thin.
4) The unit Reynolds number effect is rather small [see Fig. 21 and Eq. (4.11a)].
5) Theoretically, 100% of the dynamic pressure can be recovered, but the distance required is infinite.
6) Aside from error in the theory, the curves of Fig. 21 are the shortest possible pressure recoveries—they are the "end of the line." Nothing better can be done except by boundary-layer control.
7) The Stratford pressure distribution is the path of least resistance connecting two pressure points A and B, as will now be shown.

If we accept quadrature formulas such as Spence's or Truckenbrodt's, then Stratford's flow is found to be the minimum-drag flow. Consider the flow situation sketched in Fig. 22. We wish to go from some point A to another point B. Various

Figure 22 Possible pressure distributions connecting points A and B.

paths may be possible, as sketched. The Stratford path is the path of lowest drag. We shall show this by means of Truckenbrodt's formula.[26] For purposes of demonstration, we can write it as follows:

$$\theta_B = \left[\theta_A + K\int_{x_A}^{x_B}\left(\frac{u_e}{u_\infty}\right)^{3.33} dx\right]^{6/7} \tag{4.12}$$

The path that developes the minimum value of the momentum defect θ_B at B will be the path of minimum drag. It is obvious by inspection that the Stratford path is the answer. For it, the Truckenbrodt-type integral is minimum. As has already been shown, any curve below Stratford's would suffer from separation. According to this method of analysis, the best path is Stratford's, even if it means a sudden jump upward in velocity as sketched at B. Exactly how this flow can be applied to airfoil design is uncertain, in general, because of all the interacting effects. But at least we should be able to say that when a Stratford flow is applied in a sound fashion, low drag can be expected.

The first part of Eq. (4.lla) of the pressure rise is quite rapid, but the final part, Eq. (4.11b), is quite slow. It is enlightening to compare with a lossless two-dimensional diffuser as sketched in Fig. 23, which provides a simple physical picture. In canonical form, (with, V as velocity) the pressure coefficient referred to conditions at x = 0 is

$$\bar{C}_p = 1 - (V^2 / V^2) \tag{4.13}$$

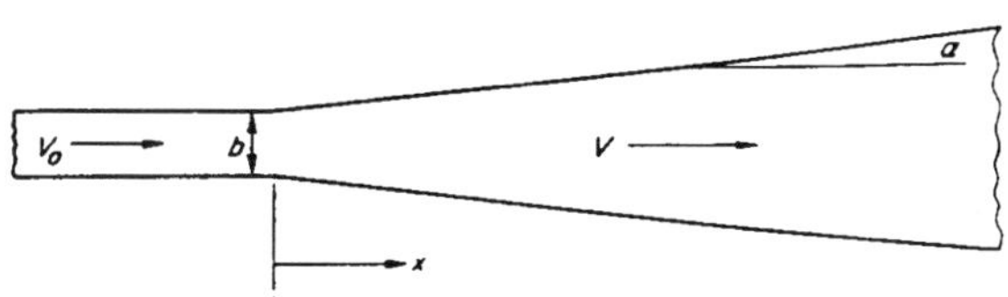

Figure 23 Two-dimensional diffuser.

The width of the diffuser is $w = b + 2x \tan \alpha$. When that expression is combined with the equation of continuity and substituted into Eq. (4.13), we have

$$\bar{C}_p = 1 - \frac{1}{[1+(2x/b)\tan\alpha]^2} \tag{4.14}$$

a formula whose structure is generally like that of Eq. (4.11 b) the sense that an infinite distance for full recovery is indicated. But the exponent of the denominator is 2 for the diffuser, as compared to 1/2 for Stratford's flow.

A Stratford flow has a continuous margin of safety over its entire length with respect to separation, whether at the beginnimg of pressure rise or far, far downstream. The margin of safety can be adjusted by changing the constant S in Eq. (4.10).

Both Eqs. (4.11b) and (4.14) show that an infinite distance is required to bring a flow in the boundary layer to complete rest. Now if a round-nosed object is placed in a stream, it generates a stagnation point. Then, if a round-nosed object like a cylinder is placed on end on the floor of a wind tunnel, the situation is such that a boundary layer is brought to rest in a finite distance, with separation sure to occur. (It is believed that the three-dimensionality of the flow does not negate that conclusion.) Airplanes often have similar intersections—wing-fuselage, for example. The inevitability of separation in such cases does not seem to be fully appreciated. The inevitable separation is one of the justifications for filleting.

4.7 Limiting Suction Lift

By means of the relations in Sec. 4.4, the Stratford canonical distributions of Fig. 21 can be transformed into idealized airfoil-upper-surface pressure distributions. A key parameter in the conversion is the effective velocity at the trailing edge, that is, the effective edge velocity to which the forward flow must decelerate when a boundary layer exists. Without the boundary layer, airfoils that ended with a finite trailing-edge angle would always have zero final velocity. The effective velocity we are talking about is really a correlating velocity. Edge velocities at the 97- or 98-% chord point are good values. Because the flow over an airfoil may be accelerated, retarded, heated, sucked, blown, reenergized, and finally discharged "overboard," the author tends to think of the boundary-layer air as going through a sort of physical process and then being "dumped" overboard. Hence, he likes to refer to the effective leaving velocity by a homely name—dumping velocity.

Observation of various pressure distributions shows that a typical effective trailing-edge velocity or "dumping velocity" is $(u_e/u_\infty)^2 = 0.8$. If we use this value, we can form two main families of limiting upper-surface pressure distributions, one for full laminar flow along the rooftop and one for full turbulent flow. Some results are presented are in Figs. 24 and 25 for a chord Reynolds number of 5×10^6. Figure 25 amounts to a direct scaling of the standard Stratford flow, because the rooftop is assumed to be entirely turbulent. In Fig. 24, the rooftops are considerably longer, because with laminar flow a

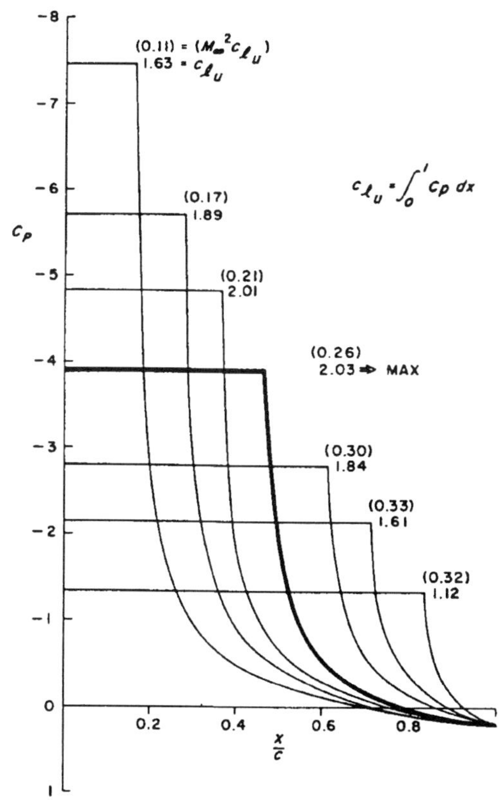

Figure 24 Suction side pressure distributions using Stratford pressure recovery to $C_p = 0.20$ at trailing-edge. Laminar rooftop, $R_c = 5 \times 10^6$. Values beside each curve indicate the lift that is developed. In the $M_\infty c_l$ form M_∞ is that value which just makes the peak velocity sonic according to the von Kármán-Tsien formula.

longer distance is required to develop the same value of R_θ as that developed by a run of fully turbulent flow.

All the flows of Figs. 24 and 25 have the property of producing the same final velocity without any separation but with the most rapid possible pressure rise from various rooftop C_p-levels. If the rooftop C_p is very high, the pressure rise must start early. If low, it can start late. If areas within the curves are calculated, it is found that a maximum exists and that the initial negative C_p for this condition is not very high. Values of the upper-surface lift coefficients are noted on the figures. For laminar and for turbulent approach flow, the maximum suction-surface lift coefficients are 2.03 and 1.00, respectively. Higher

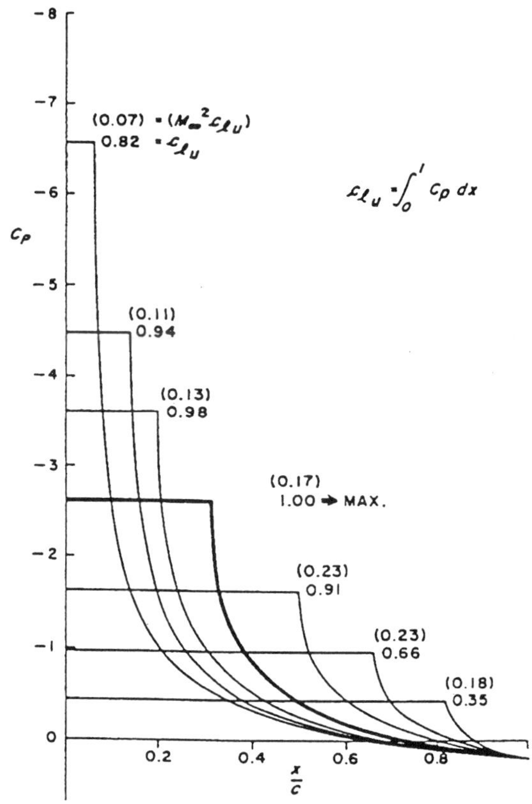

Figure 25 Suction side pressure distributions using Stratford pressure recovery to C_p = 0.20 at trailing edge. Turbulent rooftop. $R_c = 5 \times 10^6$.

Reynolds numbers would increase these values. It is interesting to note the very strong theoretical effect of laminar flow on lift. Usually, we think that the main benefit of laminar flow is reduction of drag.

We have been discussing lift coefficient analysis. A second and perhaps more important subject is the maximum force or $M_\infty^2 c_l$, that we talked about in Sec. 3. The problem cannot now be solved as accurately as for incompressible flow by means of Stratford's analysis. A compressible Stratford-type solution is necessary as a minimum. Nevertheless, nothing prevents us from determining critical Mach numbers for each of the pressure distributions by Kármán and Tsien's rule and assuming that the pressure distributions are still valid. If so, we get the values of $M_\infty^2 c_l$ in parentheses in Figs. 24 and 25. Again there is a minimum value. For laminar flow it is 0.33. The maximum occurs at much lower lift coefficients.

In the two figures, the y-ordinate was scaled to produce $(u_e / u_\infty)^2 = 0.8$ at the trailing edge. Suppose that we could by some means raise the number to 1.2. Then by Eq. (4.4) all values would be increased 50% and a 50% gain in lift would result. Therefore, a high dumping velocity is very important.

4.8 Examples

Within the accuracy of the theory, we have just shown in an idealized form the maximum suction lift that can be developed with fully attached flow when the flow is that in a natural boundary layer. R. H. Liebeck[27] was greatly responsible for the basic studies. He has recently applied them to actual airfoil design, where lower-surface pressure, upper-surface pressure, and shape must all be considered. Furthermore, he formulated the problem exactly, working in terms of surface distance and circulation instead of pressures projected on a plane, as in Figs. 24 and 25. The formulation leads to an inverse airfoil design problem that can be solved with great accuracy by James' method[28] or other good inverse methods.

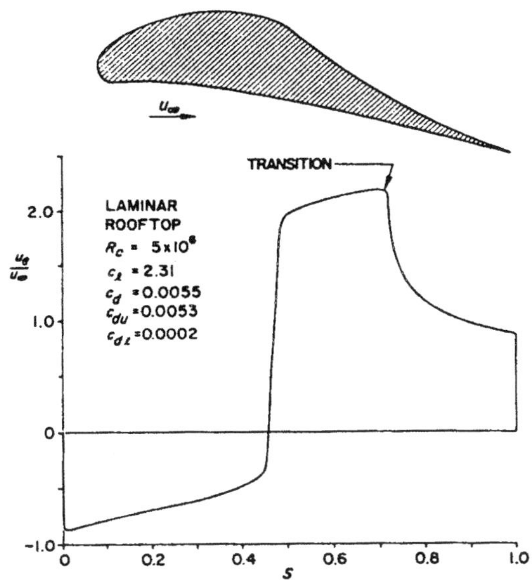

Figure 26 An optimized high-lift Liebeck airfoil that uses a Stratford pressure recovery. Values noted are theoretical. S is peripheral distance measured from the trailing edge. Pat. Pend.

One such solution is given in Fig. 26. To be sure of developing laminar flow on the upper surface, the forward part was given a favorable gradient instead of just a constant velocity flow, as in Figs. 24 and 25. Theoretical values of c_l and c_d are shown on the figure. The section L/D is 420. Tests, Ref. 7, of two airfoils of the type shown in Fig. 26 show almost perfect agreement between theory and experiment at the design point. One of the most remarkable features of the airfoil is the extremely low theoretical drag of the lower surface. As can be seen from the pressure distribution, the velocity on the lower surface is not only low but is accelerating continuously from nose to trailing edge. Hence, it is likely to be laminar all the way. The upper-surface drag is relatively low too, partly because of the minimum-drag property of Stratford flows (Sec. 4.6, property 7).

Liebeck's work was begun in an attempt to answer the question, "What is the maximum lift that can be developed by a single-element airfoil with fully attached flow, and what is the shape required to develop it?" At $R_c = 5 \times 10^6$, for all practical purposes, Fig. 27 gives the answer. The airfoil has substantially zero thickness, because at design conditions thickness is harmful, as a little thought will indicate. This limiting airfoil develops a section L/D of 600! Conventional airfoils rarely exceed 180.

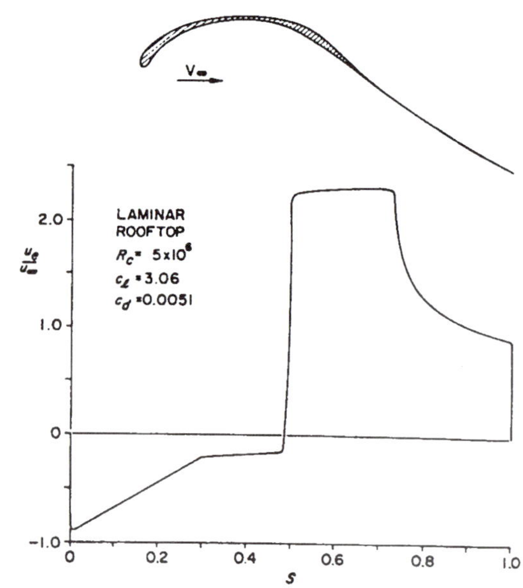

Figure 27
The Liebeck airfoil that produces maximum lift with fully attached flow, design $R_c = 5 \times 10^6$, Pat. Pend.

Although they are not related to the high-lift problem, we shall close this section with some interesting examples of what can be accomplished by various kinds of boundary-layer flows. By showing some limiting shapes, Fig. 28 demonstrates the ability of various types of boundary-layer flows to withstand adverse pressure gradients. All three shapes are cusped in order to eliminate the rear stagnation point that theoretically accompanies a finite-angle trailing edge. All boundary-layer calculations were made for the naked shapes. A slight relief would be provided if the effects of boundary layer were taken into account. Hence, the bodies shown here have a slight degree of conservatism. The first cases represent the thickest symmetrical Joukowski strut that can have complete laminar flow without any laminar separation. It is only 4.6% thick. The Reynolds numbers must be low; otherwise the flow would become turbulent. The intermediate case is a Joukowski strut having mostly turbulent flow with transition as shown in the figure. The chord Reynolds number is 10^7. Finally, if we pull out all stops and so shape the forward gradients as to be favorable to laminar flow and then use the Stratford pressure recovery in the rear, we obtain the thickest shape, which is more than 53% thick. The theoretical Michel transition

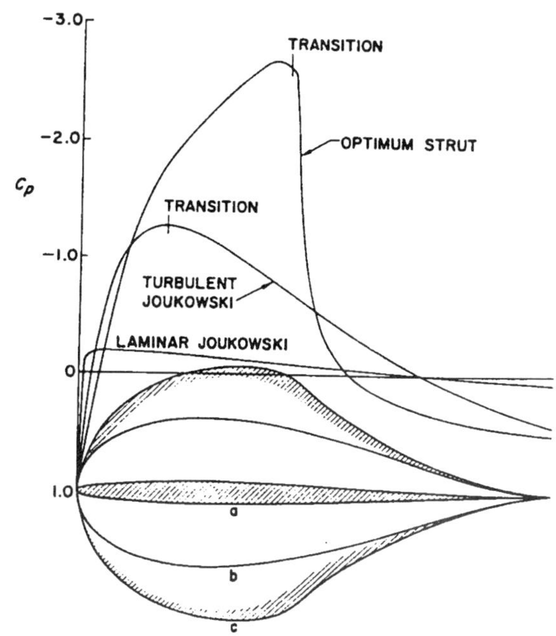

Figure 28 The thickest struts having fully attached flow.

point is marked on the figure. With the strong favorable pressure gradients, there should be no trouble in obtaining the required laminar flow, even at the rather high Reynolds number of 10^7. Like the high-lift shapes, it too was designed by the James method. Because of the extensive laminar flow, the theoretical value of c_{dp} is only 0.0077, in spite of the great thickness.

4.9 Off-the Surface Pressure Recovery

Until now, our concern has been with fluid flowing into regions of higher pressure while it is in contact with a surface, that is, a decelerating boundary-layer flow. But there can be another kind of flow, the flow of wakes that may be out of contact with any wall, into regions of higher pressure. Such a flow occurs, to a degree, on any multielement airfoil. A slat develops its own boundary layer, which flows off its trailing edge, forming a wake of low-energy air that now flows alongside the rest of the airfoil and on downstream. Consider the geometry of Fig. 17, for instance, or the streamline pattern of Fig. 29. Each forward element produces wake components over its downstream partners.

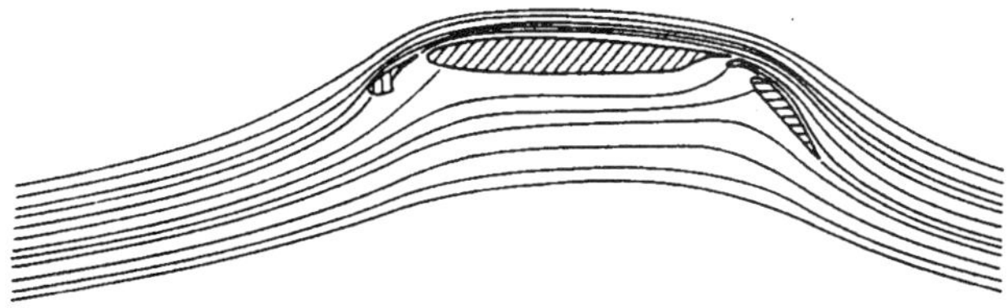

Figure 29 Calculated streamline flow field for airfoil with leading edge slat and double slotted flap, $\alpha = 0°$, $c_l = 3.70$.

The theory of that kind of wake flow is not nearly so well developed as the theory of boundary-layer flow. Therefore, we shall be content to give only a brief discussion of its features, chiefly for the purpose of calling attention to them. There can be two kinds of wakes flowing into a region of higher pressure. One is separated from the adjacent boundary layer by a region of potential flow. That kind occurs when gaps are large. The other kind is so close to the adjacent boundary layer that the two flows finally merge and become one thicker boundary layer. That kind occurs when gaps are small. By some it is called confluent boundary-layer flow.

Because a wake is usually near the main airfoil surface, the pressures impressed on it are little different from those on the airfoil surface; for example, consider the streamline in Fig. 29 that leaves the slat. If the pressure rise is great enough, we can have flow reversal in the stream, entirely off the surface. The phenomenon is easily demonstrated by resorting to Bernoulli's equation. Consider a flow as illustrated in Fig. 30. At station 0, there is a flow in which static pressure is constant across the stream, but there is a wake-like portion in which velocities are deficient. The wake

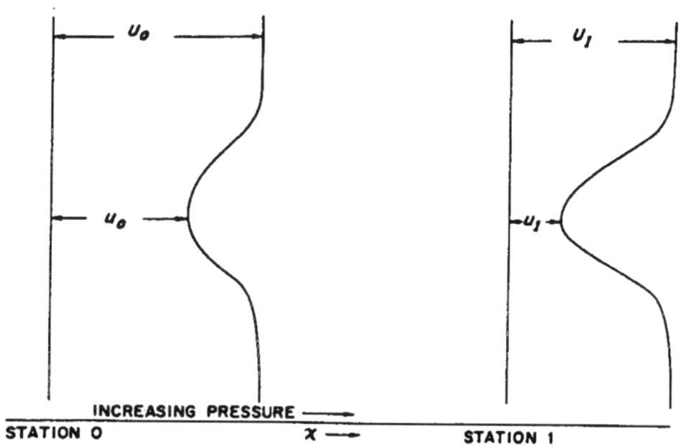

Figure 30 Flow of a wake into a pressure rise.

flows into a region of higher pressure at station 1. What happens to the wake region? Because we are considering regions very thin with respect to any overall curvature of the flow, it is reasonable to assume that the static pressure p_1 at station 1 is also constant across the boundary layer. Because gradients $\partial u/\partial y$ will be small, the shear stresses will be low and then it is a good approximation to assume that each streamline maintains its total head between station 0 and station 1. Such an assumption has been confirmed in numerous analyses, including Stratford's. Hence, using U for potential regions and u for energy deficient regions, we can write, for incompressible flow,

$$p_o + \tfrac{1}{2}\rho U_o^2 = p_1 + \tfrac{1}{2}\,\rho U_1 \tag{4.15}$$

and

$$p_o + \tfrac{1}{2}\rho u_o^2 = p_1 + \tfrac{1}{2}\,\rho u_1 \tag{4.16}$$

Solve for U_1 and u_1 and obtain their ratio, thus:

$$\frac{U_1^2}{u_1^2} = \frac{p_o - p_1 + 1/2\, pu_o^2}{p_o - p_1 + 1/2\, pU_o^2} \tag{4.17}$$

Now introduce our canonical pressure coefficient $\bar{C}_p = (p - p_o)/(1/2\,)\rho U_0^2$ into (4.17). We obtain

$$\frac{u_1^2}{U_1^2} = \frac{(u_o^2/U_o^2) - \bar{C}_p}{1 - \bar{C}_p} \tag{4.18}$$

The equation shows that as $\bar{C}_p$ increases, u_1^2/U_1^2 can reach zero well before $\bar{C}_p$ reaches + 1, provided that $u_o^2/U_o^2 < 1$. For example, if $u_o^2/U_o^2 = 1/2$, $u_1^2/U_1^2 = 0$ when $\bar{C}_p = 1/2$. That is, the velocity-defect ratio is magnified, and flow reversal can occur in the main stream. Viscosity, of course, helps to smooth out the wake velocity defect, which means that a Bernoulli approach is unduly conservative.

According to Eq. (4.18), if flow is into a region of higher pressure, the velocity defect ratio always worsens. Opposing it is the effect of viscosity, which tends to smooth out the defect. Gartshore[29] has derived an approximate test as to whether the wake grows or decays. It is

$$\frac{1}{1 - C_p}\frac{dC_p}{dx} = \frac{1}{1 - \bar{C}_p}\frac{d\bar{C}_p}{dx} > \frac{0.007}{\delta^*} \tag{4.19}$$

where δ^* is the displacement thickness of the wake at the point being considered. If the left-hand side of Eq. (4.19) is less than $0.007/\delta^*$, the wake decays; if greater, it grows.

To study the structure of Eq. (4.19), we shall apply the test to a flow that is a variation of the $\bar{C}_p = x^m$ flows shown in Fig. 20. Referring to that figure, we shall assume that we have flat-plate flows of various lengths, from 0.015625 to 1.0 ft, that develop boundary layer on both sides. All end at $x = 0$. They then dump their boundary layers into the $\bar{C}_p = x^m$ pressure region. Will the wake defect amplify or decay? First we must know the initial displacement thickness of the wake. Assume it to be equal to the final displacement thickness of the boundary layer from both sides of the plate. A convenient formula is Blasius 1/7-power formula

$$\delta^*_{1\ \text{side}} = \frac{0.0463l}{R_l^{1/5}} \tag{4.20}$$

After substitution from (4.20) into (4.19), we obtain

$$\frac{1}{1-\overline{C}_p}\;\frac{d\overline{C}_p}{dx} > \frac{0.075R_t^{1/5}}{l} \tag{4.21}$$

where l is the length of the flat plate and $R_l = u_0 l/v$. Figure 31 shows the results. The two sides of Eq. (4.21) are plotted separately. When the left-hand side exceeds the right-hand side, instability develops. The $m = 1$ curve is a special dividing case. It begins with the value of 1.0. Any higher value of m begins with the value 0; any lower value of m, the value ∞. Hence, according to Gartshore's criterion, the wake is always initially unstable for m <1.0. The figure shows a strong effect of wake thickness upon the stability. For our problem a higher Reynolds number is favorable because it reduces δ^* and hence increases the value of the right-hand side of Eqs. (4.19) or (4.21), but the effect is weak. According to Eq. (4.21), for the $\overline{C}_p = x^m$ flows the left-hand side ultimately approches infinity, and therefore any of the wake flows should finally go unstable. Hence, with $m <$ 1, theoretically we have a situation where the wake flow is unstable to begin with, but quickly becomes stable and finally becomes unstable again if carried far enough. Whether that indeed is true is not known.

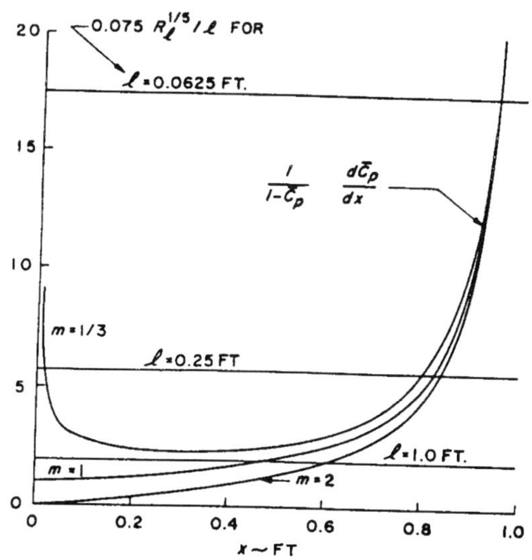

Fig. 31 Evaluation of Eq. (4.21) for the decelerating flows of Fig. 20, $u_0/v= 10^7$

However, according to Gartshore's analysis, a boundary layer is more prone to separation than the wake is to instability. Consider the l = 0.25-ft plate in Fig. 31. For any of the m-values, the final instability is at $x \simeq 0.8$ ft. However according to Fig. 20, for most values of m, the boundary-layer flow has separated earlier.

Gartshore cites experimental evidence that confirms his deductions. It was obtained by examining flow over deflected plain flaps. Without doubt, the effect can and has occurred—separation off the surface. But since there has been very little study of that kind of flow, the validity of relation (4.21) is in question. Especially in question are some of the derived consequences, such as the initial instability for some fractional values of m. For example, when m= 1/3 and 1/2, a boundary-layer flow would not separate, but the wake flow would be unstable. The principal purpose in presenting example applications is to get the concept into the open, and at least exhibit in a qualitative way the effect and the interaction between wake thickness and pressure gradient.

In practical applications, with their lack of infinite adverse gradients through which a wake must flow, the wake-instability problem should rarely be critical, which means that wakes can endure pressure rises that boundary layers cannot endure. What a multielement airfoil does then, in effect, is to use two methods of pressure recovery: the conventional, that is, on-the-surface pressure recovery, and off-the-surface pressure recovery. On a slatted airfoil, for instance, the history of the flow is this: Air flows over the slat, reaches a peak velocity, then decelerates in contact with its surface, leaves the surface and continues to decelerate until trailing- edge pressures are reached, after which it gradually accelerates back to freestream conditions. By the off-the-surface deceleration, recovery from very high negative C_p values can be made in much shorter distance than can be made when all the deceleration is in contact with a surface.

On the right-hand side of Fig. 17 is a canonical pressure scale based on the maximum velocity on the slat. According to that scale, the final velocity-squared ratio at the trailing edge is only 0.025. If all the deceleration had been in contact with the surface, the flow would surely have separated, according to Figs. 19 and 20. Nothing so far has been said about the case when boundary layers merge. Lockheed[30,31[has studied the problem and, in fact, has developed a general method for analyzing confluent boundary layers. But because of the complicated nature of the flow, bold simplifications had to be made. Like Gartshore's analysis, the methods need development and further checks by detailed experimental work. Improved methods for both the merging and the nonmerging wake flows are problems for future work.

It is our observation, as well as that of Foster et al.[32], that gaps between airfoil elements should be so large that wakes and boundary layers do not merge for, if they do, early separation will set in. In fact, Foster et al. in discussing the subject makes the following statement:

"...when the flap is in the optimum position from the viewpoint of obtaining the highest maximum lift, the interaction between the wing wake and the flap boundary layer is comparatively mild, with the two layers retaining their separate identity almost to the flap trailing edge."

Hence, we are somewhat fortunate. Optimum designs are outside the region of merging boundary layers, and, furthermore, flow retardation rates are rarely so great that the wake becomes unstable. The boundary layer on an element beneath the wake is nearly always in worse trouble. However, the merging of boundary layers helps establish the optimum. If the boundary layer were much thinner, an optimum spacing might be less. Hence, because of scale effects on the boundary layer, the optimum spacing for an aircraft may be appreciably less than indicated by a small scale wind-tunnel model. The problem is worst for the slat because its wake has the longest run.

5. Loads Applied to Boundary Layers by Airfoils

5.1 Introduction

The problem of obtaining high lift is that of developing the lift in the presence of boundary layers—getting all the lift that is possible without causing separation. Provided that boundary-layer control is not used, our only means of obtaining higher lift is to modify the geometry of the airfoil. Considerations that guide the modification, then, are the subject of this section. It is helpful to think in terms analogous to those of structural design—applied and allowable loads. The airfoil applies the loads, and the boundary layer determines the allowable. Separation of a boundary layer may be likened to reaching the yield point in materials testing. Initial separation rarely coincides with the maximum lift that an airfoil develops, for the lift usually continues to increase. Likewise, the yield point in a metal is not the point of ultimate load. That usually is the rupture point. Hence, the point of maximum lift coefficient may be likened crudely to the rupture point of a material. The analogy is quite rough, of course, but it is mentioned because the interaction between boundary layer and shaping of an airfoil is not very widely appreciated. We close these introductory remarks by observing that aerodynamic science has advanced to the point where we can satisfactorily predict the point of initial separation (the yield point) but not the condition of maximum lift (the rupture point). The second problem is still beyond the state of the art, but it is assuming high priority because of successful solution of the simpler problems.

5.2 Single-Element Airfoils

The subject of single-element pressure distribution types and consequent airfoil performance has been indirectly covered by the material in Sec. 4. The canonical pressure distributions show general limits to the pressure rises, and the pair of charts can be used for visual checks prior to careful examination of more nearly final designs by detailed boundary-layer calculations. If separation is indicated for the problem at hand, some change must be made in the shaping in order to remedy the defect. In making the change, it must be borne in mind that definite bounds on what can be accomplished have already been established. For a one-piece airfoil, there are several possible means for improvement—changed leading-edge radius, a flap, changed camber, a nose flap, a variable-camber leading edge, and changes in detail shape of a pressure distribution. A pressure rise may be improved by changing from convex to concave in the direction of Stratford's type. Even changing the trailing-edge angle from finite (e.g., 10°) to cusp may be useful because a cusped trailing edge imposes less pressure rise at the trailing edge. Furthermore, a cusp increases the slope of the lift curve, so that a given C_L is reached at a lower angle of attack, which lowers nose pressure peaks.

Simple hinge systems as on a plain flap, even though sealed, can have a significant adverse effect on the separation point and hence on lift. Figure 32 shows the adverse effects of the break at a hinge line. The airfoil is an NACA 63A010 airfoil hinged on the lower surface at the 75%-chord point. Also shown for comparison is a variable-camber flap whose centerline is a circular arc. Its final slope is 25°, so that both trailing edges have the same final angle. Both flaps have separation, but the variable-camber shape has attached flow over 85% of the flap, while the simple flap has attached flow over only 71%. It is interesting that the simple hinged flap has a concave pressure distribution and the cambered a convex. In the canonical sense the plain flap pressure recovery is more efficient.

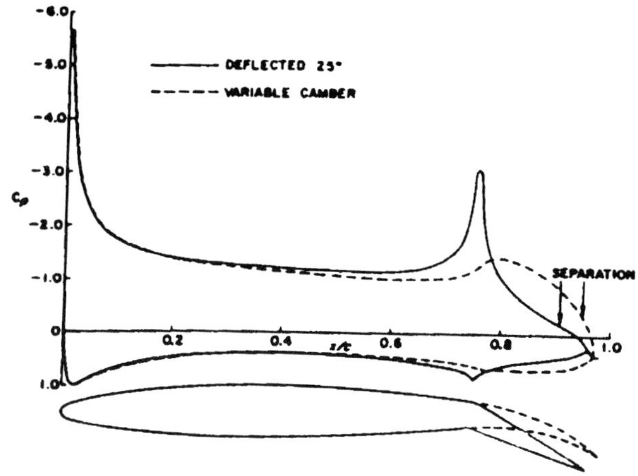

Figure 32 Comparison of two kinds of flaps on a NACA 63A010 airfoil. For the plain flap $\alpha = 0°$ and inviscid $c_l = 1.78$. The airfoil with variable camber flap was set at $\alpha = 1.06°$ in order to obtain the same c_l. Separation points are marked by arrows. $R_c = 10^7$. Transition is at forward suction peak.

The variable camber flap turns out to be better only because of its drastic reduction in the suction peak.

Canonical pressure distributions are useful for preliminary scanning. Figure 33 shows a conventional pressure distribution. Where is separation most imminent? Near the front on top? Near the rear on top? Or on the bottom? Figure 34 is the canonical form of Fig. 33. Comparison with the curves of Figs. 19 and 20 indicates that the nose is far from separation. The rear upper surface is marginal, but only behind about 97% chord. The lower surface retards to about $\bar{C}_p$ =0.5 in a rather short distance, and the retardation begins late. Figures 19 and 20 indicate that a $\bar{C}_p$ -value of about 0.6 could be reached in this distance. Hence there seems to be some margin, but not much. Provided that other constaints would allow it, the faintly S-shaped pressure distribution aft of 50% chord on the bottom could have been improved by replacing it with one that is concave all the way. In problems and approaches like this, the need for inverse methods become evident. First, we would like to have an inverse boundary- layer method that for arbitrary initial conditions could tell us what pressure distribution we could have that contains a certain margin against separation. Then, given the specified pressure distribution, the method should find the airfoil shape that produces it. Definite progress is being made along these lines.

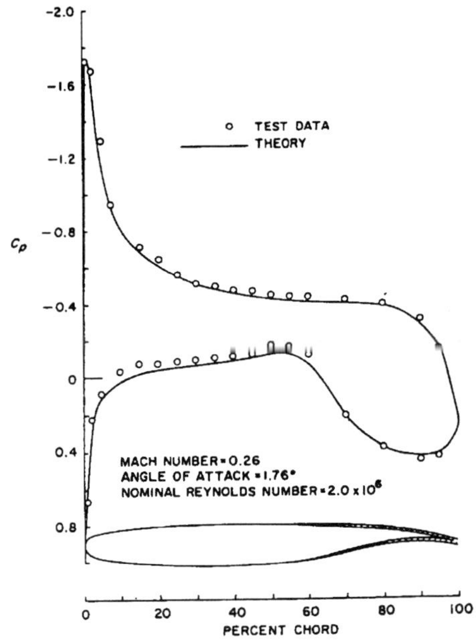

Figure 33
Pressure distribution for a typical aft-loaded airfoil. Pressure distribution is corrected for boundary-layer effects. The shaded area on the airfoil represents the displacement thickness.

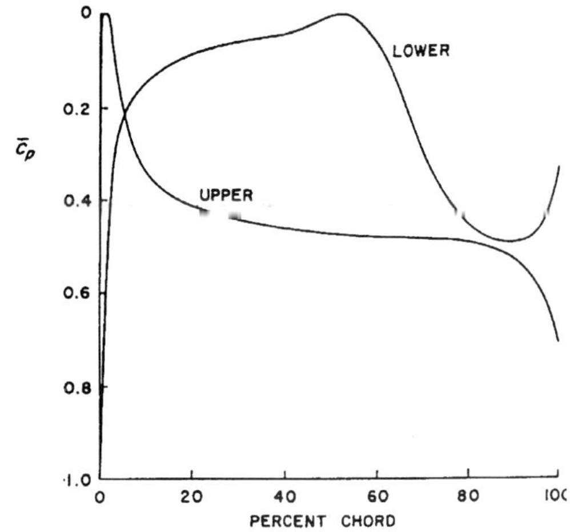

Figure 34
Pressure distribution of Fig. 33 in canonical form.

5.3 Multi-element Airfoils—General

There seems to be a great deal of ignorance and confusion about the effect of gaps in properly designed multielement airfoil systems. One misconception was mentioned in the introduction to Sec. 4. The most common misconception is that a slot supplies a blowing type of boundary-layer control. The idea can be traced at least as far back as Prandtl,[33] who said, "The air coming out of a slot blows into the boundary layer on the top of the wing and imparts fresh momentum to the particles in it, which have been slowed down by the action of viscosity. Owing to this help the particles are able to reach the sharp rear edge without breaking away." Abbott and von Doenhoff[34] merely make the safe comment, "Slots to permit the passage of high energy air from the lower surface to control the boundary layer on the upper surface are common features of many high-lift devices." Here again, boundary- layer control is implied. Perkins and Hage[35] make the harmless and noninformative statement: "The air flowing through the slot in Fig. 2-49 [sic] is accelerated and moves toward the rear of the airfoil section before slowing down and separating from the surface." Lindfield, in Lachmann,[7] has a brief paper on the slot effect. Part of the article describes studies by Lachmann, in 1923, who represented a lifting slat by several vortices located near a circle. The circle was then transformed into a Joukow-ski airfoil. Lachmann's theoretical studies seem to have been largely forgotten and not really appreciated. However, he considered only half the problem—the effect of a slat on an airfoil. He did not consider the effect of the airfoil on the slat. Lindfield's article is correct as far as it goes, but it is not very factual. He points out the need for better analytical methods before the slot effect can be well analyzed. (Now 14 years later, most of the methods are available.)

A remark about the action of slots comes from a recent NASA report.[30] "It is well recognized that the usual function of the slot is that of a boundary-layer control device permitting highly adverse upper surface pressure gradients to be sustained without incurring severe separation. This stabilizing influence results from the injection of the high energy slot flow into the upper surface boundary layer." A still more recent NASA report[36] states, "This leading-edge slat gives the fluid which passes through the gap between the slat and the main airfoil a high velocity. Consequently, a boundary layer which grows on the upper surface of the main airfoil has more momentum than it would have in the absence of the slat."

There are two things wrong with these statements. First of all, the slat does not give the air in the slot high velocity. If anything, it gives the air low velocity. Secondly, the air through the slot cannot really be called high-energy air. All the air outside the actual boundary layers has the same total head. Properly designed and spaced slats are far enough apart that each component develops its own boundary layer under the influence of the main stream, and there is no merging within the slot. Topologically, the process of boundary-layer development is no different from that on a biplane. Subject to their particular pressure distributions, the two boundary layers on a biplane grow, trail off downstream, diffuse, and finally merge. That is just the process for an airfoil system of two or more elements so long as merging does not occur within the slot.

The next paragraphs will elaborate on and confirm what we have just said. There appear to be five primary effects of gaps, and here we speak of properly designed aerodynamic slots.

1) Slat effect—in the vicinity of the leading edge of a downstream element, the velocities due to circulation on a forward element, for example, a slat, run counter to the velocities on the downstream element and so reduce pressure peaks on the downstream element.

2) Circulation effect—in turn, the downstream element causes the trailing edge of the adjacent upstream element to be in a region of high velocity that is inclined to the mean line at the rear of the forward element. Such flow inclination induces considerably greater circulation on the forward element.

3) Dumping effect—because the trailing edge of a forward element is in a region of velocity appreciably higher than freestream, the boundary layer "dumps" at a high velocity. The higher discharge velocity relieves the pressure rise impressed on the boundary layer, thus alleviating separation problems or permitting increased lift.

4) Off-the-surface pressure recovery—the boundary layer from forward elements is dumped at velocities appreciably higher than freestream. The final deceleration to freestream velocity is done in an efficient manner. The deceleration of the wake occurs out of contact with a wall. Such a method is more effective than the best possible deceleration in contact with a wall.

5) Fresh-boundary-layer effect—each new element starts out with a fresh boundary layer at its leading edge. Thin boundary layers can withstand stronger adverse gradients than thick ones.

Those effects will now be explained and discussed in turn at some length. Laminar bubbles, merging boundary layers, and the like may complicate the problem; but when Reynolds numbers are high and at design conditions, such side effects should not be important. Therefore, only conventional boundary-layer effects are considered.

5.4 Slat Effect

Figure 35 displays the slat effect. A slat that is lifting has circulation in the direction sketched in the upper part of the figure. To a first approximation, a slat may be simulated by a point vortex. A more accurate approximation would account for thickness by means of several sources and sinks, sources in front and sinks behind, the total yield being zero, of course. For our purposes, it is sufficient to consider only the vortex. From the sketch it is evident that the velocities induced on the airfoil by the vortex run counter to those that would develop on the nose of the isolated airfoil, especially at high angle of attack. Pressure peaks would therefore be reduced. Liebeck and Smyth[37] studied the effect by

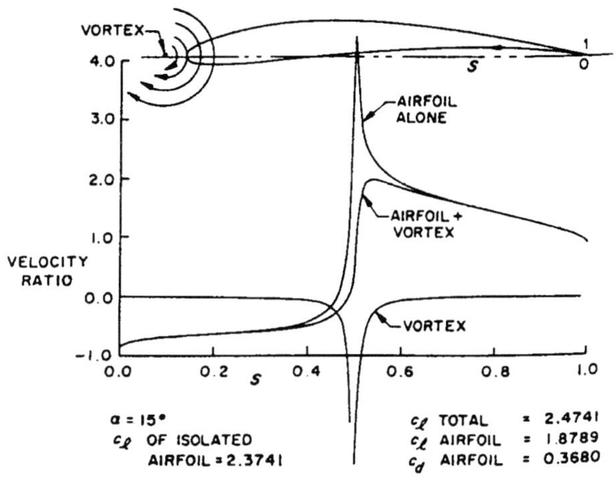

Figure 35 Velocity distributions on an airfoil with and without a vortex located as shown. *S* is the arc length around the airfoil surface beginning at the trailing edge, measured in a clockwise direction. Total perimeter is unity.

means of a special generalized Joukowski airfoil program combined with computer graphics. Figure 35 is an example of the work. In the method, the vortex could be moved around at will. The best location found for the vortex was typical of the location of a slat. The "best" location is not a precise statement. It is the location where the nose suction peak was largely eliminated and resulting pressure distribution was smooth. At $\alpha = 15^\circ$, the plain air-foil developed a very strong suction peak. As is noted at the bottom of the figure, c_l for the airfoil alone is 2.37. With a vortex located as shown, the pressure peak was completely eliminated, the airfoil c_l fell to 1.88, but the total lift, including the vortex, increased to 2.47. In Fig. 35, the curve called "vortex" shows the velocity induced on the airfoil by the vortex in the absence of the main stream. For best cancellation of the isolated airfoil suction peak, vortex position and strength were so adjusted as to generate a negative velocity peak that was roughly the mirror image of the isolated-airfoil suction peak. Flow near the rear of the airfoil was almost unaffected, as can be seen by examination of either the airfoil pressure distribution or of velocities induced by the vortex.

That then is the slat effect. Contrary to the implication in the various quotations cited in Sec. 5.3, the velocity on the nose of the airfoil is reduced. Peak nose velocity ratio was reduced from about 4.4 to under 2.0. Obviously, the boundary layer is much better able to negotiate the modified distribution. With a vortex operating, there is only a very small increase in total lift, a fact that is consistent with wind- tunnel observations. With slat extended, the main effect is to delay the angle of stall, not particulary to shift the angle of zero lift. In the beginning of this discussion, it was mentioned that thickness effects could be accounted for approximately by sources in front and sinks behind, to approximate the shape of the flap. Since total sink strength must equal total source strength and since the sinks are closest to the airfoil, the net effect of the combination is the induce a small forward velocity component on the airfoil. Hence thickness effects should generally reinforce the circulation effects.

There is an interesting and more physical way of explaining the slat effect. Again consider Fig. 35. When the airfoil is at an angle of attack, flow whips rapidly around its nose, which has a small radius. High centrifugal forces are developed. Without outside help, high negative pressures around the nose are needed to balance out the centrifugal force. But the vortex is a turning aid, and a true slat also would be one. There should be a close correlation between the amount of load carried by the slat and the average C_p - reduction over the nose. The author looked at Fig. 35 and made the following crude estimates: Area modulated = 10% chord, mean C_p of the airfoil alone for this region = - 8. Mean C_p for same region with vortex present = – 2. Net C_p - change = 6. Then approximate lift force c_l that must be supplied by the vortex is 6 x 0.6. The actual lift on the vortex is the difference between c_{ltotal} and $c_{lairfoil}$. According to the numbers at bottom of Fig. 35, c_l = 0.6 is indeed the load carried by the vortex. Probably, the analysis cannot be made quantitative, but it does add to the understanding of the flow process. The problem would be a good one for further study. The force exerted by the vortex is not just vertical; it has a drag component too. According to the figure, because net drag must be zero, its c_d-value is -0.3680, a negative force that causes slats to extend automatically. A more careful analysis would consider this component.

The slat effect on a real airfoil is illustrated well by Fig. 36, which represents a well-developed, practical design. The pressure distribution for the airfoil without slat is also shown. The quantity $(u_e / u_\infty)^2$ reached a maximum value of 10.04 (C_p = – 9.04). With the slat in place, the value was reduced remarkably, to just over 3. The effect is strikingly similar to that of the point vortex in Fig. 35. The figure also shows the upper-surface pressure distribution in canonical form. Again, signs of a jet blowing effect are conspicuous by their absence. This and most of the other pressure distributions were calculated

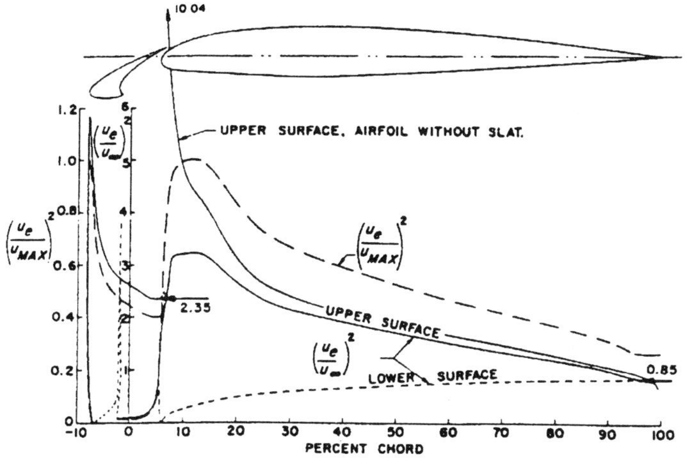

Figure 36 Typical theoretical pressure distributions for a two-element airfoil, corrected for the boundary layer. The dumping velocity is denoted by the arrows. Also included is the canonical pressure distribution for the upper surface. as well as pressure distribution for the main airfoil alone. The latter exhibits clearly the favorable effect of a slat. $\alpha = 13.15°$, $c_l = 1.45$ for the two-element airfoil.

by one version or another of the Douglas Neumann program[38,39] that has been tested by so many hundreds of comparisons with wind-tunnel and other data that its predictions can be accepted as being essentially exact, at least for our current purposes.

Figure 37 was specially synthesized for purposes of this paper. It consists of three identical airfoils, all at the same angle of attack. For comparison and indication of the degree of interference, the pressure distribution for the isolated airfoil is included. Observe the extreme rounding of the peaks of the two rear airfoils, which again is the slat effect. A small amount of the rounding is undoubtedly due to the decreased velocities that accompany a converging flow, as at a trailing edge. As in the point-vortex example, the rear of each airfoil is covered with a surface sink distribution, according to the Neumann program. The surface sinks induce a counter-flow over the following airfoil. Although it was not done, it would be easy to learn the thickness effect by replacing a forward element by a much thinner section while keeping the same mean line.

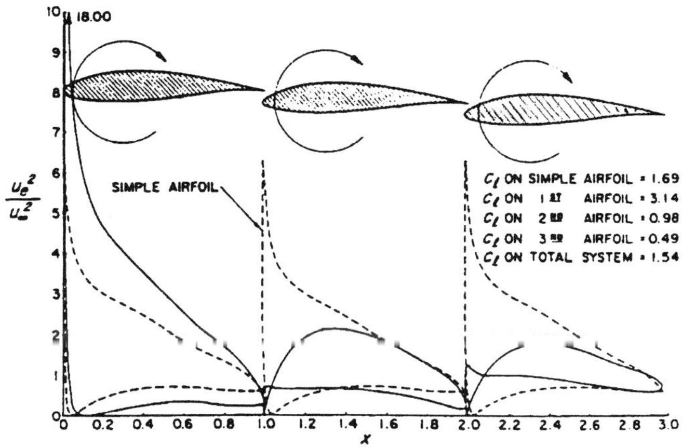

Figure 37 Pressure distribution on a three-element airfoil formed from three NACA 63_2-615 sections, arranged as shown. All are at the same angle of attack, 10°. Shown also is the pressure distribution on the basic simple airfoil at 10° of attack. Slot gaps are 1% of each chord.

5.5 Circulation Effect

The slat effect has been recognized before; in fact, it was clearly recognized as far back as 1923 by Lachmann.[7] But the circulation effect does not seem to be explicitly recognized. Figure 38, which shows the effect in its simplest form, was produced by the computer graphic setup used in Fig. 35. Since the vortex could be placed anywhere in the field, it was now placed near the trailing edge, as indicated in the figure. If it were to represent the lift of a slotted flap, the vortex circulation would be as indicated. The flow effectively places the trailing edge at a high angle of attack, and if the Kutta condition is to be met, the circulation must increase. The effect is little different from that of deflecting a small plain flap on the isolated airfoil. In that case, the onset flow would be approaching the rear of the airfoil at a considerable angle. But for Fig. 38 we did not turn the trailing edge; instead, we used a device to turn the flow. Figure 38 shows a slightly different airfoil from that in Fig. 35, but again the angle of attack is the same. The vortex has a drastic effect on circulation, increasing c_l from 2.59 to 3.49. In Fig. 38, the vortex-alone curve shows the velocity distribution around the airfoil that is induced by the vortex. That distribution does not meet the Kutta condition, but the airfoil vortex combination does, of course. It is obvious from the figure that any means of changing the flowfield near the trailing edge would change the lift. Hence, a source, properly positioned,

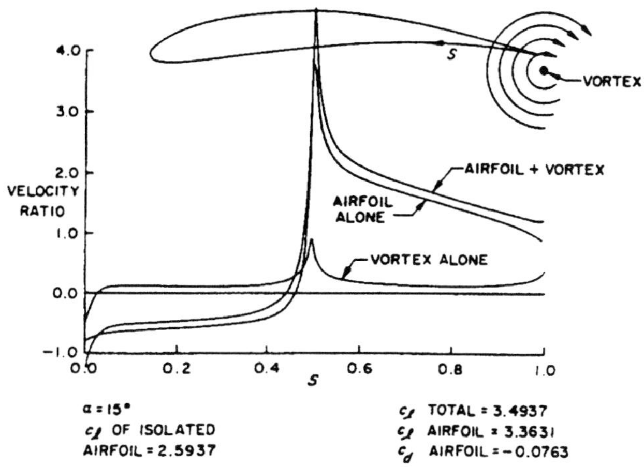

Figure 38 Point vortex used to simulate a slotted flap. Vortex increases c_l of airfoil at $\alpha = 15°$ from 2.59 to 3.36. Leading edge pressures are greatly increased by action of the vortex.

is apt to be as effective as a vortex. The effects on circulation that are due to aft distortions of the flow are then what we call the circulation effect.

According to Fig. 38, the final upper-surface velocity is increased considerably over the upper-surface velocity for the isolated airfoil. Because of the additional velocity caused by adding the vortex to the general translational flow, the airfoil discharges boundary layer at the trailing edge into a stream that is locally of higher velocity. That is the "dumping effect" that has already been mentioned and that will be discussed further in Sec. 5.6.

We have just indicated that any method capable of introducing cross flow at the trailing edge will influence the circulation. An obstruction, properly placed, can be a powerful factor for controlling the circulation. To illustrate such control, we used a circular cylinder, which is about as neutral a body as can be selected. Figure 39 shows the system studied. Two circular cylinders of different diameters are centered on a ray from the trailing edge. The gap is constant at 0.1 chord. The angle of the ray, δ, was varied from 0° to 90°. The lift on the airfoil and on the airfoil-circle combination was calculated. Figure 40 shows some results for the case where the cylinders were directly to the rear of the chord line ($\delta = 0°$). Large increases in lift are indicated. The cylinder itself carries a large amount of lift, even though it has no trailing edge. Figure 41 shows the effects in more detail, as functions of δ. As might be expected, an optimum deflection angle is found. The most effective angle is about 60° or 70°.

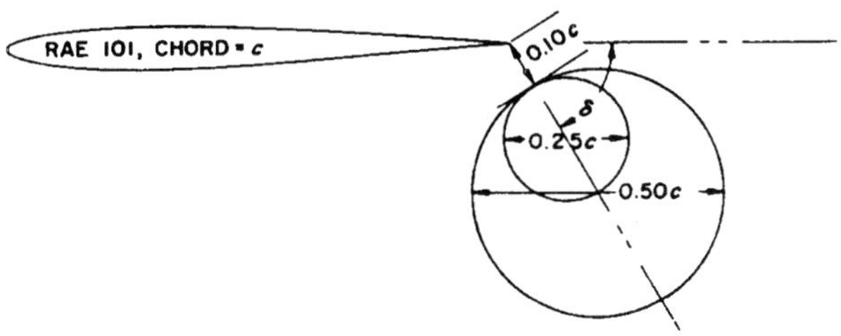

Figure 39 Airfoil-circular-cylinder combinations studied to learn effect of an obstruction on circulation.

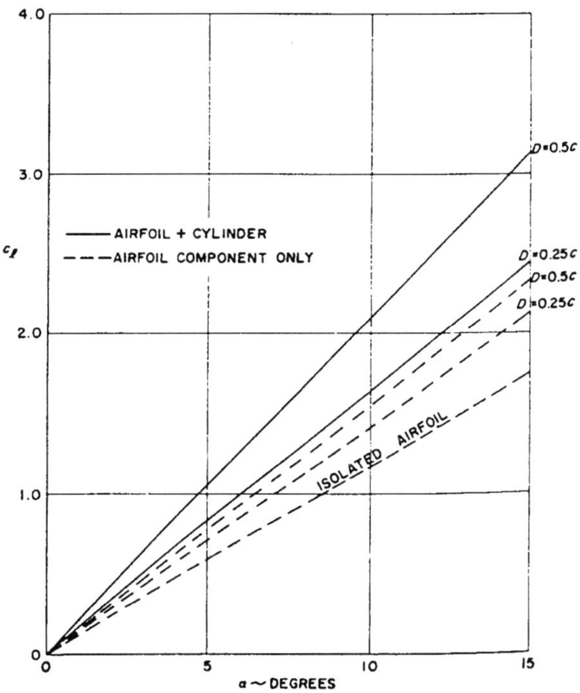

Figure 40 c_l vs α curves for airfoil-cylinder combinations, showing strong effects on circulation. Deflection δ of cylinder = 0°.

At 15° angle of attack, the lift coefficient of the isolated airfoil is c_l = 1.75. With the 0.50c circular cylinder set at δ = 60°, c_l= 3.35 for just the airfoil, nearly double the value for airfoil alone. Hence the effect is very great. Figure 39 is drawn with δ = 60°. It appears that this most effective position is similar to the position found most effective for slotted flaps and slats in combination with airfoils. Considered as a control surface, the control effectiveness $\partial c_l/\partial\delta$ is not much less than that of a plain flap.

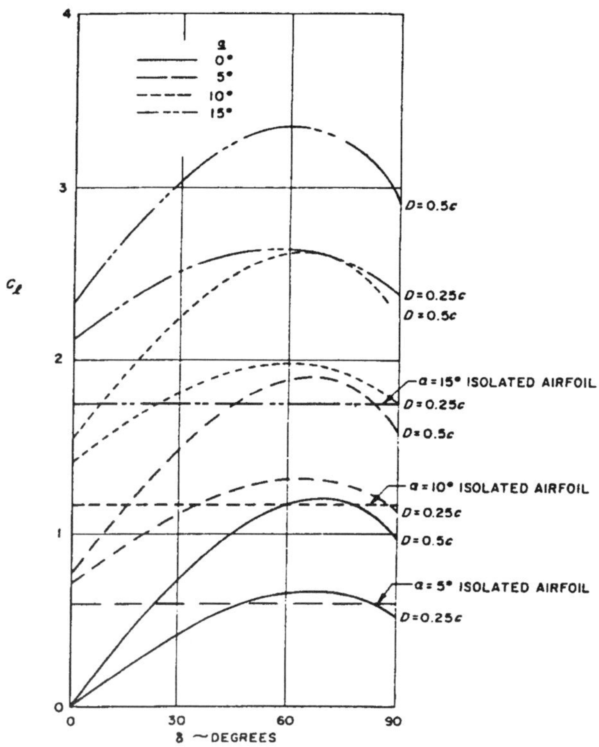

Figure 41 Airfoil-circular-cylinder combinations. Effect of α_o, δ, and diameter on lift coefficient.

Figure 42 shows calculated pressure distributions for the cases with δ = 60° and with c_l held constant at 1.5. The corresponding values of α are noted in the figure. Peak velocities at the nose are considerably reduced by the cylinder. If the pressures were plotted in canonical form, the ones with the cylinder would appear considerably more favorable. Velocity ratios at 0.975c, which amount to the dumping velocity, are tabulated. The cylinders double the dumping velocity ratio. However, according to Figs. 19 and 20, separation should still occur in all cases. That is not surprising, in view of the specified c_l and shape of the airfoil.

The velocity u_θ on the surface of an isolated circular cylinder in crossflow is $u_e / u_\infty = 2 \sin \beta$, where β is angle from the nose. For our problem, the velocity component that is normal to the airfoil is $2 \sin \beta \cos \beta = \sin 2\beta$. The maximum value of that component, which occurs at $\beta = 45°$, is not much less than the $\delta = 60°$ maximum seen in Fig. 41. The rough coincidence justifies the assumption that the change in circulation is a function of the cross flow induced at the trailing edge by other bodies.

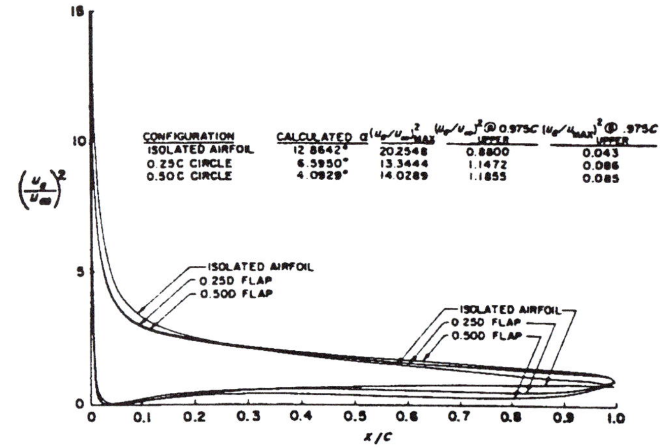

Figure 42 Pressure distribution at $c_l = 1.5$ for airfoil-cylinder combinations. $\delta = 60°$.

Another aid in thinking about the problem of modifying the circulation is to consider the rear stagnation point in nonlifting flow. Without circulation, the rear stagnation point is on the upper surfacc at some distance forward of the trailing edge. The farther it is moved forward, by whatever means, the greater the circulation required to move it back to the trailing edge. The two guidelines—cross-flow strength and nonlifting stagnation-point location are the best that can be proposed as means of understanding and designing to maximize the circulation. The factors cannot be isolated—there are too many interactions. Fortunately, accurate multi-element-airfoil analysis methods are available, together with much experience. But perhaps factors such as those discussed here will be of value in endeavors to improve design. Rarely do people involved in design of a slotted flap understand why the best place to position the nose of the flap is slightly behind and below the trailing edge of the main airfoil. They just know that with the flap in that position the system worked. We hope we have explained why.

The reason why augmented circulation is helpful needs to be given. In numerical airfoil methods of the Douglas-Neumann type, the complete solution is a linear combination of three fundamental solutions, which are shown in Fig. 43. Because of the small radius of the leading edge, local velocity ratios in that region are very high indeed for both the 90° and the circulation solutions. A positive lifting circulation is shown in Fig. 43c. It is clear that its disturbance velocities add to those of the 90° solution for positive angles of attack. For an ordinary airfoil, increasing the lift requires increasing the angle of attack. But in the process more of the undesirable 90° solution enters into the summation. Now if the airfoil nose can be kept at low angle of attack, a gain will be made if circulation can be produced by other methods. A plain flap increases that circulation because when it is deflected, a considerable cross-flow component of the onset velocity exists at the trailing edge. The vortex or an obstacle at the trailing edge has the same effect. To avoid nose peaks, the nose should be at a slight negative angle in any real design, in order that the velocities in the circulation and in the 90° solutions oppose each other and cancel out, or more precisely, the linear combination of the three solutions substantially cancels out. In short, considering the separation problem, high lift is best obtained by keeping the nose angle of attack low and inducing circulation by means other than pitching.

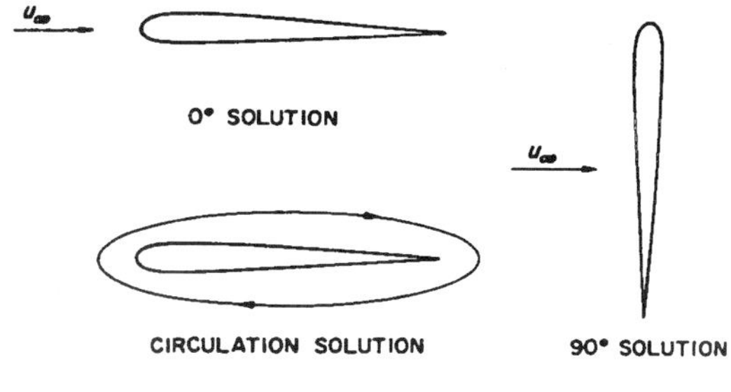

Figure 43 Three fundamental solutions used for calculating flow about a lifting airfoil at angle of attack.

Figure 44 illustrates the circulation effect. The c_l on the rear airfoil, which has no device to augment its circulation, is only 0.49. The rear airfoil boosts the circulation of the middle airfoil, which has a c_l of 0.98. The two rear airfoils act

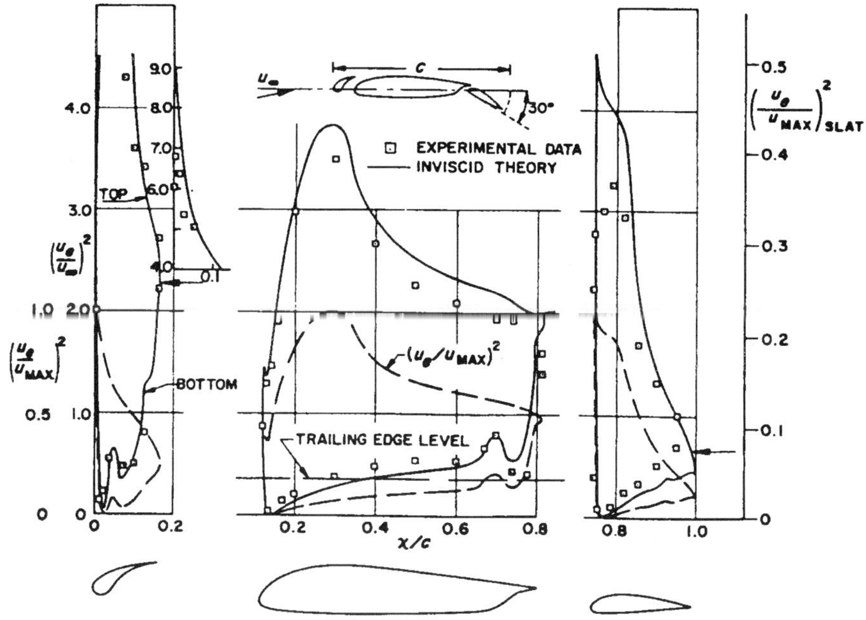

Figure 44 Typical three-element airfoil, showing dumping velocity effect. Arrows denote dumping velocity. NACA 23012 airfoil, $\alpha = 8°$. Line marked trailing edge level is trailing edge dumping velocity divided by maximum velocity on main element.

together on the front airfoil, giving it a c_l of 3.14. There is some feedback, of course. Circulation on a forward element effectively reduces the angle of attack of a rear element, as well as the velocities on its surface, and hence reduces its lift. That is one reason for the very low lift of the rear element.

In Sec. 4.1, it was suggested that slots amount to a short- circuiting of the lift. That effects seems to exist if the elements are not properly positioned. Observe the drop in pressure near the slots in Fig. 44. It is interesting to note that at 10° angle of attack, according to inviscid-flow calculations, the single-element airfoil carries slightly more load than the three-element system.

The same kind of interaction occurs between sails on a sailboat or even between sailboats in close proximity. According to Fig. 44, if the airfoils correspond to sailboats, the last two boats, which are drawing up on the leader, would find it very difficult to pass him. They are augmenting his lift and reducing theirs. The effects are even stronger between sails on the same boat. A useful and very interesting study of the problem has been made by A. E.

Gentry.[40] By using the electrical-analogy technique and conducting paper, he explained correctly for the first time the jib-mainsail interaction effect.

5.6 Dumping Effect

Closely related to the circulation effect is the dumping effect. The favorable interference of a downstream element induces cross flow at the trailing edge that enhances circulation of the upstream element. But the interference may also increase velocities in a tangential direction so that the flow from a forward element is discharged into a higher velocity region, thus reducing pressure-recovery demands. The effect is quite favorable to the boundary layer. According to Eq. (4.4), the suction lift can be increased in proportion to $(u_{T.E.})^2$ for the same margins against separation.

Does the effect really exist? Yes, indeed! It can be seen in any properly designed multielement airfoil. It is clearly shown theoretically in Fig. 42; values of the dumping-velocity ratio are tabulated in the right-hand columns of the table in the figure. For the three cases—plain airfoil, airfoil plus 0.25*c* circle, and airfoil plus 0.50*c* circle—the velocity ratios are 0.88, 1.147, and 1.186, respectively. The canonical ratio nearly doubles; it increases from 0.043 to 0.085.

Figure 36 is a theoretical case for an airfoil with slat. For the main airfoil, the dumping-velocity ratio squared is 0.85. On the slat, it is much higher: 2.35. The pressure distributions are also shown in canonical form, and according to them the slat is less severely loaded than the main airfoil—in the sense of margin against boundary-layer separation.

Figure 44 shows experimental data for a three-element airfoil. Again the dumping-velocity effect is clearly displayed, this time confirmed by experiment. For the three surfaces, starting with the flap, the dumping-velocity-squared ratios $(u_e/u_\infty)^2$ are 0.67, 2.0, and 2.28. The canonical plots indicate that, for the conditions shown, the main airfoil is less severely loaded than the slat and the flap. The enhanced dumping velocity shown is typical. The pressure distribution of any properly designed multielement airfoil shows it; one only need look for it.

It is interesting to conjecture what might be done with the effect if an inverse design method for multielement airfoils became available, assuming that the complete airfoil requirements permitted. Start with some desired pressure distribution for a hypothetical three-element airfoil, for example, that of Fig. 45. The basic pressure distribution that we shall use is shown as the dashed line.

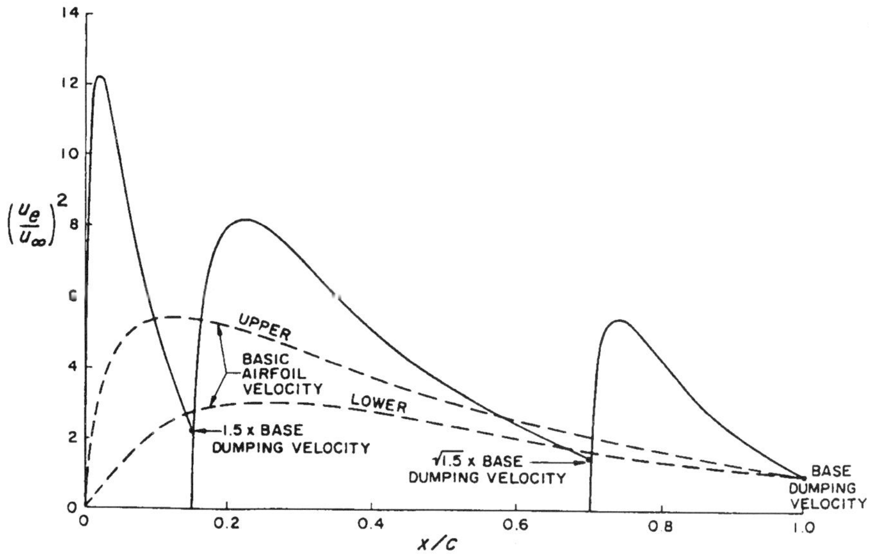

Figure 45 Illustration of compounding of lift by using multielement airfoils. For this illustration the factor *m* is (1.5) ½

Then suppose we seek a three-element airfoil that has that canonical pressure distribution for all three elements. Furthermore, assume that the rear element induces a dumping-velocity~squared ratio $(u_e/u_\infty)^2$ of 1.5 on the middle element. Because that element now has higher circulation, it can induce even greater effects on the front element. Assume the induced velocity ratio is the same for each element with respect to its own velocities, which is a bold assumption. With that assumption, the pressure distribution for the three elements in Fig. 45 is as sketched. The load carried is far greater than that of the equivalent single-element airfoil, yet the several canonical pressure distributions are all the same.

The effect, as just described, is readily quantified. It will be done only for a three-element airfoil, but the analysis is easily generalized. Let c = chord of the ensemble; $f_1 c$ = chord of rear airfoil (element 1); $f_2 c$ = chord of center airfoil (element 2); and $f_3 c$ = chord of front airfoil (element 3). Then for the three elements $f_1 + f_2 + f_3 = 1$. Also let m_2 = magnification ratio for velocity at trailing edge of element 2 due to being in the high-velocity field of the nose of element 1; and m_3 = magnification ratio for velocity at trailing edge of element 3 due to being in the high-velocity field of the nose of element 2.

Let the upper-surface lift coefficient be c_{l_0} for the given basic pressure distribution. Then for the rear element the suction-side lift is $f_1\ c_{l_0}$. On element 2, the wind speed over the whole element is greater by a factor m_2. Hence for element 2, the lift is $f_2\ c_{l_0}\ m_2$. For the front element, the apparent wind speed is increased by a factor m_3. But the speed on the middle element has already been increased by the factor m_2. Hence on element 3 the lift is $f_3(m_2m_3)^2\ c_{l_0}$. Then the lift of the ensemble is

$$c_l = f_l\ c_{l_0} + f_2 m_2^2\ c_{l_0} + f_3(m_2m_3)^2\ c_{l_0} \tag{5.1}$$

and the ratio is

$$(\ c_l / \ c_{l_0}\ ) = f_1 + f_2 m_2^2 + f_3(m_2m_3)^2 \tag{5.2}$$

For the example of Fig. 45, $f_1 = 0.30$, $f_2 = 0.55$, and $f_3 = 0.15$. Also, $m_2 = m_3 = (1.5)^{1/2}$. Then, according to (5.2),

$$(\ c_l / \ c_{l_0}\ ) = 0.30 + 1.5\ (0.55) + 2.25\ (0.15) = 1.463$$

that is, a 46% gain, with no more tendency for separation than on the base airfoil. The effect definitely exists, but (5.1) is as much hypothesis as it is statement of fact. In any real airfoil design, so many more factors enter into the problem that is is not nearly that simple. Compressibility problems would ultimately set in, for instance.

5.7 Off-the-Surface Pressure Recovery

Off-the-surface pressure recovery has already been discussed in Sec. 4.9. Without its assistance, boundary layers would often be unable to meet all pressure-rise demands. Consider Fig. 44, which represents theory roughly confirmed by test data. The scale on the right-hand side shows the total deceleration ratio from the peak velocity. At the trailing edge of the flap, the velocity-squared ratio has decreased to 0.075. Figure 17 shows a theoretical pressure distribution that can be realized experimentally. Its final velocity-squared ratio is 0.025. According to Figs. 19 and 20, the flow would never have reached such very low velocity ratios if it had been a simple boundary-layer flow that was always in contact with a wall. But in multielement flows, the front element's boundary layer, which undergoes the greatest deceleration ratio, is largely decelerating off the surface. Figure 4 illustates an air foil that shows a theoretical total deceleration ratio $(u_{T.E.}/u_{max})^2$ equal to 0.0045.

It is opportune to mention something that has been almost taken for granted—that any accurate calculation of pressure distributions must take into account the displacement thickness of the boundary layer. That assumption is well covered in Ref. 39. In this paper, we have not been consistent, because consistency is not necessary to illustrate our point. In general, our simple theoretical calculations do not consider boundary layers. In other cases, we note in the caption whether or not boundary-layer corrections were included.

A wake alongside a body has only a slight effect on the inviscid theoretical lift. If a wake has a square profile, it can be approximated by sheets of vorticity, positive vorticity on one side and negative on the other. Obviously, the wake has some effect on circulation, although the two sheets almost cancel each other. Furthermore, if the wake flow is curved, there is a centrifugal force of the same sort that is developed by the curved sheet of a jet flap, except that here the momentum amounts to a defect instead of an excess. Foster, Ashill, and Williams,[32] and Ashill[41] have carefully examined those effects in both theoretical and experimental studies. They develop formulas for the various effects. In general, they found that, for properly designed multielement airfoils, the only effect of importance is that of the displacement thickness. Moser and Shollen-berger[42] also have studied the problem theoretically and experimentally. They found that the theoretical inviscid effect of the wake is to reduce lift by a small amount, typically a few hundredths in c_l. Their theory is in satisfactory agreement with experiment. In one particular study that used an NACA 4415 airfoil, they found that c_l was reduced by an amount 2.3 c_{ds}, where c_{ds} is the drag coefficient of the slat. All the findings have, in general, been confirmed by practical experience.

To summarize—without the displacement-thickness correction the errors are greater than desirable. With the displacement thickness considered, the errors are so small that a design engineer would hardly know what to do with more accuracy—in the case of fully attached flow, of course.

5.8 Fresh-Boundary-Layer Effect

On each element of a properly designed multielement airfoil, the boundary layer starts out afresh. It is well known, of course, that thinner boundary layers can sustain greater pressure gradients than thicker ones. That fact is demonstrated well by Figs. 19 and 20. It can be displayed in a more analytic form by Stratford's formula, Eq. (4.10), which we here repeat.

$$\bar{C}_p \left[x \frac{d\bar{C}_p}{dx}\right]^{1/2} = (10^{-6} R)^{1/10} S \tag{4.10}$$

where S is a constant. The term $(10^{-6}R)^{1/10}$ varies so slowly that it is almost constant. But x, a measure of distance from the origin of the flow, enters to the 1/2 power, and, obviously, longer boundary layers admit less adverse gradient. Viewed in another way, we see that if x is halved, $d\bar{C}_p/dx$ can be doubled with the same safety against separation. Therefore, breaking up a flow into several short boundary- layer runs is favorable with respect to the delay of separation and hence to the increase of lift. It might be argued that breaking up the surface into a number of elements steepens gradients just as much as length is shortened, leaving $x(d\bar{C}_p/dx)$ unchanged. In view of the favorable interference and shape of the total pressure distributions, the quantity does indeed reduce.

5.9 Are Two Elements Better Than One?

Our previous discussion indicates that several elements are better than one element. But there has been no proof. Here we shall attempt a proof. The proof will be for a two-element case, but by the nature of the development it will be seen that the proof could also show that three elements are better than two, four better than three, etc. We make the following assumptions: 1) the lower surface plays only a weak role, and hence only the upper surface needs to be considered; 2) Reynolds number effects are weak—which they are at large scale. Hence the lift on an airfoil whose chord is unity is just the same as the lift on two of the same airfoils whose chords are each 1/2; 3) by some means, perhaps an inverse method, the elements can be so shaped that both develop the same canonical upper-surface pressure distribution. The shape of the pressure distribution is not restricted; and 4) boundary layers and wakes of forward elements do not degrade the lifting capacity of following elements.

Consider Fig. 46. Let ABCDE be pressure distribution on the upper surface of an airfoil of chord AE. That same load distribution could be applied twice, to two 1/2- chord elements as loads ABC'D'E" and E"B"C"DE. The sum of those two areas equals area ABCDE. By proper inverse methods, the shapes of two elements that could generate the same pressure distribution can probably be

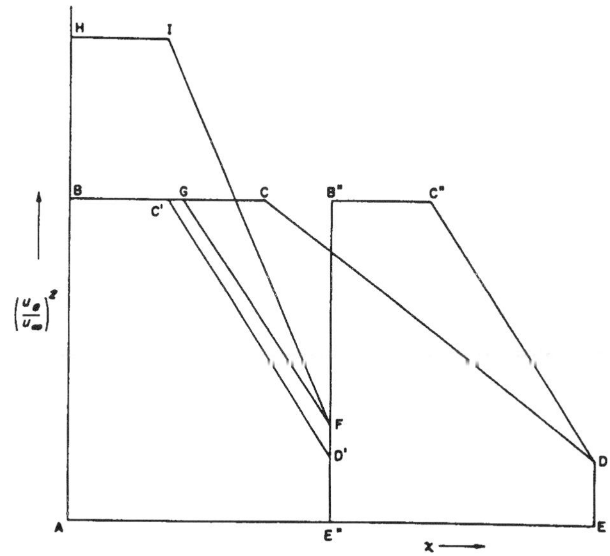

Figure 46 Load diagram for proof that a two-element airfoil can develop more lift than a single-element airfoil.

found, although that is a weak point of our proof. For the two half-size pressure distributions, the basic dumping velocities squared are E"D' and ED, which are equal. Now on any properly designed multielement airfoil, the flow of the trailing edge of the forward element can be made to discharge into the high-velocity region of the rear element. Hence for the front element, the dumping velocity squared is increased from E"D′ to E"F. The increase is the key to our proof. Two cases now arise:

a) the maximum velocity has some kind of limit; and

b) the maximum velocity has no limit.

Consider case (a) first. Perhaps the velocity limit comes from Mach number considerations. Because E"F<E:D', the start of pressure rise can be moved back from point C′ to a new point G. Now, area ABGFE" >ABC'D'E", which proves our case. Consider case (b). If dumping velocity is increased from E"D′ to E"F, the entire forward pressure distribution can be scaled in the ratio E"F/E"D′ to obtain AHIFE." Again, and even more strongly, area AHIFE" >ABC′D′E". Q.E.D.

We have not assumed that the basic canonical pressure distribution be any kind of optimum—just that it be carrying its capacity load without separation. It is not necessary to the proof that the two elements be of equal chord. One possible weakness of the proof is that for a two-element airfoil the final velocity DE cannot be maintained at as great a value as it can for a single-element shape, which means that all pressure levels would be depressed. That is a problem that can best be answered by inverse multielement-airfoil theory, for which there is a great need and a great future.

An attempt to apply these concepts has been made using Wilkinson's approximate multielement inverse method.[43] By means of the method we searched for the shape of a two- element airfoil that had the same thickness distribution and the same canonical upper surface pressure distribution as in Fig. 26. The distributions were to apply to both elements and the dumping velocity ratio of the forward element was to be $(2)^{1/2}$ times that of the rear element so that all u_e values on the front element would be twice those on the rear. Chords of the two airfoils were to be equal. Figure 47 is the result. A drastic change in shape is shown. We obtain a c_l-value of 3.82, compared with 2.31 for the single element airfoil. Because both pressures are identical in the canonical sense, there is no greater danger of separation on the two-element airfoil than on the single-element airfoil. Only one gap size and position was tried. Other gap geometries would change the shape. The method cannot be called entirely satisfactory; partly because of iteration difficulties. It is noticed that the asked for pressure distribution is measurably different from the best that could be found using Wilkinson's method. Whether the asked for pressure distribution is even possible

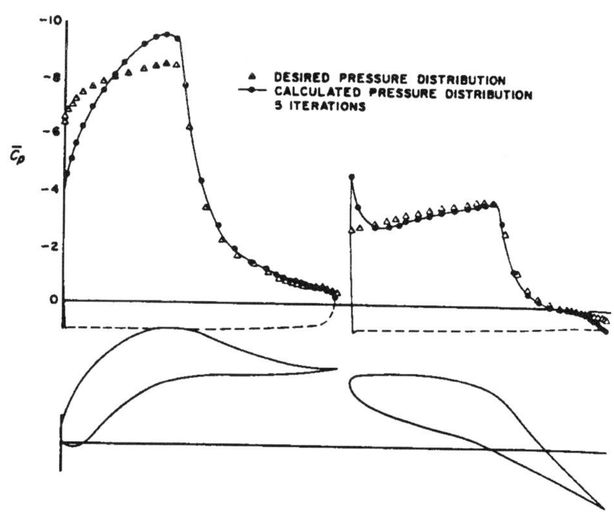

Figure 47 Shapes producing Liebeck-Stratford pressure distributions similar to that of Fig. 26. Velocity-squared values on the forward element are required to be magnified by a factor of two. c_l of the combination is 3.82.

to develop of course is unknown. It is well known that every shape has a corresponding pressure distribution, but not every pressure distribution has a corresponding shape. Nevertheless, the result is indicative of what can be done and of things to expect in the future. It confirms our proof that two elements are better than one.

5.10 Two Spectacular Examples of High-Lift Airfoils

In Sec. 2, Handley Page's 8-element airfoil system was shown (Fig. 4). It occurred to me that it might be of interest to calculate its theoretical pressure distribution, not only to show the power of modern computation but also because it might afford information about interactions, dumping velocity, and other features not clearly exhibited in the usual multielement airfoils that are so strongly hardware oriented; where they are influenced by the requirements of construction and restriction.

Coordinates were measured from a photographic blowup of the illustration in Handley Page's original paper.[5] (It is hoped that his original drawing was accurate.) Figure 4 shows the results. His 8-element shape developed a wing $C_L \simeq 3.3$ at $\alpha = 42^\circ$ (see Fig. 5). The wing was rectangular and had an aspect ratio of 6. According to wing theory, the midspan section was at an effective angle of attack of about 36°, the value used in our two-dimensional calculations. No boundary-layer corrections were made. The total calculated section c_l was 4.33. Velocity ratio instead of velocity ratio squared was used in order to avoid excessively large ordinates on two of the figures (C_p less than -36 was indicated).

Values of c_l in each velocity diagram represent the lift carried by that element, referred to full chord. Four of the diagrams show a second pressure distribution. It is the one that would develop if the airfoil was alone, at the same angle of attack. The leading element is at a considerable negative angle of attack, with a strong pressure peak on the lower side, which might cause separation. Nevertheless, the element is very highly loaded. In spite of the increasing angle of attack of each element, the lift contribution decreases continuously towards the rear. The final element carries a c_l of only 0.08, in spite of being at almost 90° angle of attack at the trailing edge. If it were in isolation, it would be carrying c_l of 6.80.

Although not converted to canonical form, none of the pressure distributions appears to be very close to separation, even that of the front element. Perhaps that is not surprising, for we are showing a condition that represents maximum lift for an airfoil whose chord Reynolds number was about

250,000, making the Reynolds number of each element about 50,000. At such low Reynolds numbers, separation should occur quite early, which makes the theoretical pressure distributions look quite safe when judged by our current high Reynolds number standards.

Figure 5 should be reexamined, now that our calculation has been made. In general, it confirms our conclusion that—if properly designed—the more slots, the better. But the seven-slot configuration is inferior to the six-slot configuration. The reason is not known, but the author suspects it is due to drastic modification of the trailing-edge contours when the eighth element was formed (see Fig. 4). Figure 48 shows the dumping-velocity ratio for each of the elements. There is a rather smooth and definitely continuous increase from rear to front. The rearmost element has a value of about 0.4, and the front element reaches 2.25.

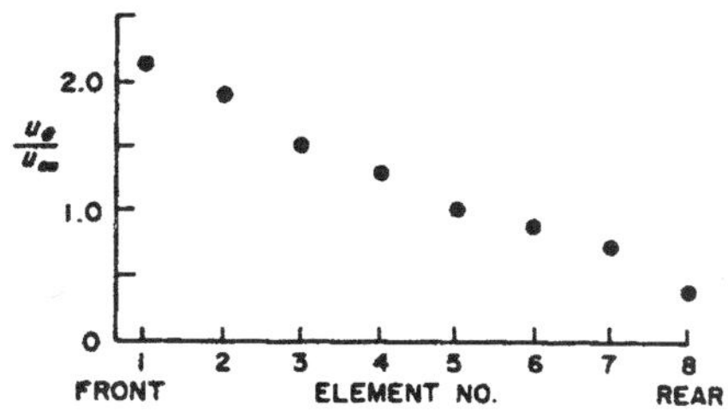

Figure 48 Upper surface dumping velocity values for the airfoil and conditions of Fig. 4. The ratio $(u_{T.E.}/u_{max})^2$ for the ensemble is 0.00456.

Figure 49 shows the extremes that can be reached in producing special-purpose high-lift airfoils. The section shown was designed by R.H. Liebeck of the Douglas Aircraft Company as a wing for Dan Gurney's Indianapolis-type race cars. The shape is a result of careful direct-method analysis and modification of his optimum airfoil studies. As installed on a racer, the wing is rectangular and of aspect ratio 2.15, with small end plates. The wing was designed theoretically, then built without benefit of wind-tunnel tests, and tested on the racer. At an angle of attack of 14°, tufts used to indicate flow separation remain fully attached and the maximum wing-section c_l is about 5, according to calibrated spring-deflection tests.

In the past few years the word "synergistic" has become popular. Those examples, the proof of Sec. 5.9, the dumping effect discussed in Sec. 5.6, as well as the various test data, all indicate that airfoils properly placed in tandem are fine examples of synergism.

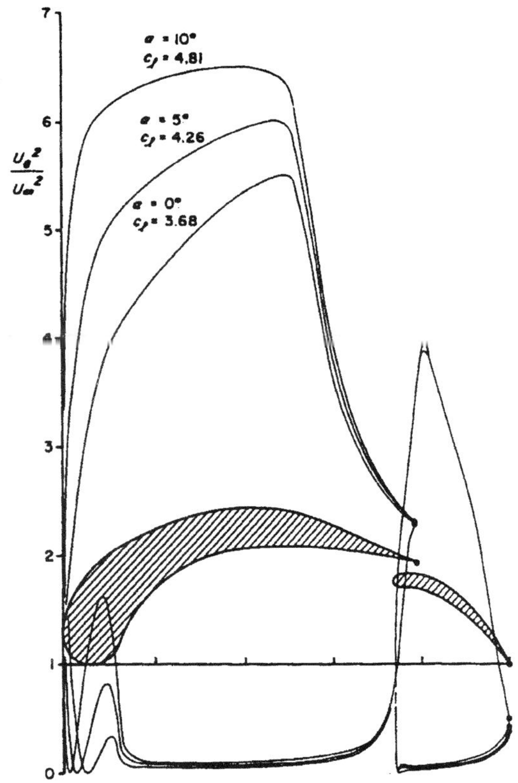

Figure 49 Special-purpose Liebeck high-lift airfoil designed for Indianapolis-type racers (Pat. Pend.).

6.0 Power-Augmented Lift

6.1 General

A few remarks should be made about the subject of power-augmented lift. Power properly applied, as through boundary-layer control, can greatly delay separation and hence increase maximum lift. Powered lift augmentation usually requires pumps and internal ducting, which amounts—at least—to weight and cost complications. There is no doubt that higher lifts can be obtained, but whether they make for a better total airplane system is a moot question. The answer depends on the efficiency of the lift-augmentation system. Here we shall confine ourselves to three studies that touch on the theoretical performance possibilities, together with ideas for improving the overall system.

6.2 Area Suction

The thorough studies reported in Ref. 44 show that turbulent flows with mass transfer can be calculated with nearly as good accuracy as turbulent flows over impervious walls. A generalized eddy-viscosity formula has been developed that is responsible for that accuracy. Therefore, it was felt appropriate here to throw some light on the subject of delaying separation, by considering the theoretical suction requirements. An illuminating procedure is to use the $\bar{C}_p = x^m$ canonical plots presented earlier, in Figs. 19 and 20. Because suction introduces another variable, space does not permit us to present results for all four lengths of flat-plate run. Instead, we consider only the one-foot run, which case may be considered roughly to approximate an airfoil with a flap.

In conventional boundary-layer calculations where a transformed stream function is used, we are given the wall conditions that $f''_w \sim u_w = 0$ (no slip) and $f_w \sim \psi_w = 0$ (impervious wall). Then the problem is to find $f''_w \sim c_f$. In the calculations to be presented, we keep $f''_w = 0$, of course, but seek values of f_w that will maintain an arbitrarily chosen constant $c_f = 0.001$. The altered procedure is just a switch in boundary conditions that is entirely acceptable according to mathematical theory. Some kind of real limiting suction would be found for the special case $c_f = 0$, as in Stratford flows, but because the machine program involves numerical solution of a partial differential equation, computing and convergence problems arise when c_f is set equal to zero. We, therefore, avoided those difficulties by specifying that suction would be sufficient to maintain c_f at a value equal to 0.001. Thus the results to be shown contain a certain margin of safety. Previous studies have specified the suction and then found the resulting skin friction. Figures 50 and 51 show results for (u_0/v)-values of 10^6 and 10^7.

The results are shown in terms of values of v_w/u_e required to maintain c_f at the chosen value 0.001. Suction starts where c_f had decreased to 0.001. At first the suction rates are quite small, but the rates increase rapidly as the higher pressure region is penetrated. The final contour $v_w/u_e = 0.05$ is one for which a variable constant in the boundary-layer equations is about to exceed the range of its validity. Furthermore, the boundary-layer equations start losing their validity when v_w is not very small compared with u_e. Total suction flow rates can be determined approximately by integrating the v_w-values. According to the figures the effect of Reynolds number is not strong.

The charts show that suction is highly effective in causing small delays in separation. But when large delays are desired, suction loses efficiency, so that

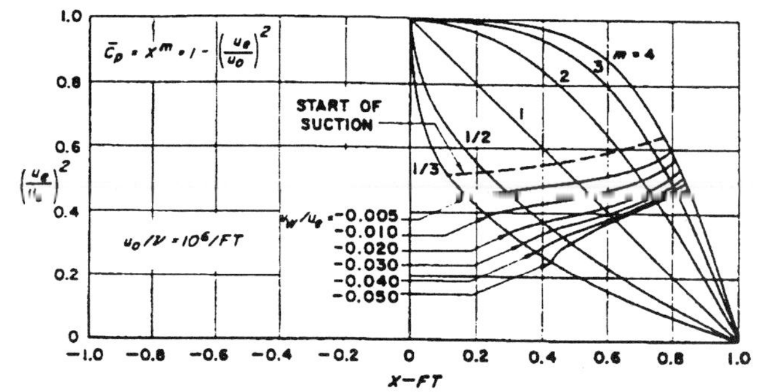

Figure 50 Effect of area suction on separation. Suction starts where c_f has decreased to 0.001 and then v_w is adjusted to maintain this value of c_f. Initial flatplate run is 1 ft, turbulent, $u_o/v = 10^6$/ft.

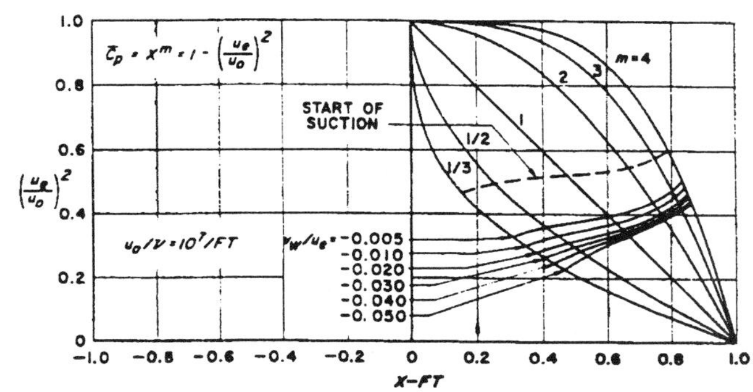

Figure 51 Effect of area suction on separation. Suction starts where c_f has decreased to 0.001 and then v_w is adjusted to maintain this value of c_f. Initial flatplate run is 1 ft, turbulent, $u_o/v = 10^7$/ft.

strong suction is required. The effect can be explained physically in a neat fashion. Assume the region of pressure rise to be quite short so that the suction can be assumed to be concentrated at a point. Let us then examine the special case of a flat-plate flow entering a sudden pressure rise assisted by suction. Long ago, in connection with Griffith airfoils, G. I. Taylor considered the same kind of problem. He first assumed that the pressure rise occurred in such a short distance that the total head along each streamline in the boundary layer would remain constant. Then he hypothesized that no filament whose total head was less than the downstream static pressure could flow into the downstream region of higher pressure. In other words, unless it was sucked off the filament would separate. These hypotheses give us a procedure for calculating the amount of suction required to prevent separation at a pressure jump. His analysis has been well confirmed by experiment for laminar flows, some of the confirmation being the author's.

Let us apply the hypotheses to our problem, namely the flows illustrated in Figs. 50 and 51, except that the pressure rise region will be approximated by an instantaneous jump. Let $(\ )_c$ designate critical conditions. Then, using our canonical flow notation we have for the total head along any streamline

$$H = p_o + \tfrac{1}{2}\rho u^2 \tag{6.1}$$

If all velocity head is expended in reaching a higher pressure, p_c, downstream, we have

$$H_c = p_c = p_o + \tfrac{1}{2}\rho u_c^2 \tag{6.2}$$

where u_c is the velocity of the dividing streamline, all fluid below being sucked off. Using $\bar{C}_p = (p - p_o)/(1/2)\rho u_o^2$, we find

$$\bar{C}_p = u_c^2/u_o^2 \tag{6.3}$$

This relation states that removal of fluid up to a height u_c will permit a canonical pressure rise from $\bar{C}_p = 0$ to $\bar{C}_{pc}$.

How much flow must be removed? Because the initial flow is flat plate it is convenient to use the 1/7 power formulas. The quantity that must be removed is $Q_c = \int_0^{y_c} u dy$ which in coefficient form is

$$c_{Q_c} \equiv \frac{Q_c}{u_o x} = \frac{1}{x}\int_0^{y_c} \frac{u}{u_o}\, dy \tag{6.4}$$

The 1/7 power formulas are

$$u/u_0 = (y/\delta)^{1/7} \text{ and } \delta(x) = 0.37x/\,(u_o x/\upsilon)^{1/5} \tag{6.5}$$

Upon insertion of the first, and integrating we have

$$c_{Q_c} = (7/8)\,(\delta/x)(y/\delta)_c^{8/7} \tag{6.6}$$

From Eq. (6.3) and the first of Eq. (6.5) we can also write

$$\bar{C}_{p_c} = (y/\delta)_c^{2/7}$$

When this, together with the second of Eq. (6.5) are substituted into (6.6) we obtain

$$c_{Q_c} = [0.324/(u_o x/\upsilon)^{1/5}]\bar{C}^4_{p_c} \qquad (6.7)$$

This formula shows that if the demanded pressure rise is indeed low the required suction c_{Q_c} is very low. If $u_o x/\upsilon$ is 10^7 then $c_{Q_c} = 0.013\ C^4_{p_c}$. If C_{p_c} = 0.5 the required c_{Q_c} is only 0.0008 but for reaching stagnation pressure it is 0.013. Unlike the results on Figs. 50 and 51 some suction is indicated for even the smallest pressure rise, but the fourth power variation shows the demand to be exceedingly small. The most interesting result from this analysis is the fourth power relation of C_Q to $\bar{C}_p$.

In Lachmann,[7] Wuest presents theoretical studies made by Pechau's approximate method. The studies were of a real suction applied to simple airfoils in various ways. Suction requirements for avoiding separation are given. Generally, our calculations show the same $C_Q = Q/u_o x$ magnitude as his, that is, of $0(10^{-3})$. We have not presented specific examples, because charts like Figs. 50 and 51 are more general. For a real engineering problem, since the boundary-layer computing method is entirely general, the particular case should be and can be studied carefully. Charts are then just a guide for preliminary estimation purposes.

The results shown represent no kind of optimum. The start of suction was arbitrarily chosen, and the intensity chosen was such as to maintain c_f at a value 0.001. A smaller c_f-value would clearly reduce v_w and C_Q. In the belief that earlier suction would be slightly less efficient, we started suction only when the boundary layer began needing help. Undoubtedly, there is an optimum distribution for each of the $C_p = x^m$ distributions, but we believe our values do not miss the optimum by very much.

There surely is some best combination of suction distribution and pressure distribution for obtaining the greatest pressure rise with the minimum C_Q. Determining that combination amounts to a calculus-of-variations problem of finding the best combination of path and suction. It is interesting to note that the basic tool for the problem has recently been developed.[45] The boundary-layer calculation method has been extended to find the pressure distribution necessary to provide a given c_f-history, which amounts to solving a nonlinear eigenvalue problem, where the eigenvalue is dp/dx. Since suction could already be analyzed by both direct and inverse methods, we now have available the tools for seeking the optimum combinations of $v_w(x)$ and $p(x)$. The problem is an interesting mathematical challenge.

6.3 Tangential Blowing

To round out the discussion, we give a limited comparison of blowing applied to the same $\bar{C}_p = x^m$ flows. The wall-jet method of Gartshore and Newman[46] is used. As for the suction case, blowing is assumed to start shortly before natural separation would occur, that is, at the $c_f = 0.001$ locus. Figures 52 and 53 are the results.

The conventional blowing coefficient C_μ is defined for two-dimensional incompressible flow as

$$C_\mu = \text{thrust}/qc = \left(\frac{2u_j^2}{u_\infty^2}\right)\frac{t}{c} \tag{6.8}$$

When applied to a family of canonical pressure distributions, both reference velocity and reference length lose their meanings. But we know, from basic boundary-layer theory and from the parabolic nature of the equations, that the direct determinants of the development of a flow subject to a certain pressure history are the characteristics of the boundary layer at the starting station. The length of prior run, the pressure distribution, and whether it is laminar or turbulent, only help to establish the boundary layer at the station where the blowing studies are started. When the boundary layer is viewed in this fundamental way, chord and freestream conditions lose their significance. Momentum thickness at the slot comes forward as the logical reference length. Furthermore, the momentum defect at this point is $\rho u^2\theta$ which is a drag force. The tangential blowing produces a thrust. We therefore introduce a new form of the incompressible blowing coefficient called C_{mq}, which is defined as

$$C_{\mu\theta} = \frac{\text{jet thrust}}{\text{momentum defect}} = \frac{\rho u_j^2 t}{\rho u_e^2\theta} = \frac{u_j^2 t}{u_e^2\theta} \tag{6.9}$$

Equations (6.8) and (6.9) can be combined to give a relation between the two blowing coefficients,

$$C_\mu = 2(u_e^2/u_\infty^2)(\theta/c)C_{\mu\theta} \quad (6.10)$$

With $C_{\mu\theta}$ as our blowing parameter, we obtain a direct "feel" for the strength of the blowing, for we are making up for the momentum lost in the boundary layer. Then, if the blowing is to be applied over only a short distance, it is to be expected that the required value of $C_{\mu\theta}$ will not be greatly different from unity.

That is indeed the case with Figs. 52 and 53. Gartshore's program deck was used for the calculations. Several different values of u_j/u_e, were tried and results were found to be sensitive to the ratio. Furthermore, since the method was only approximate, hangups were encountered in some cases when the wall-

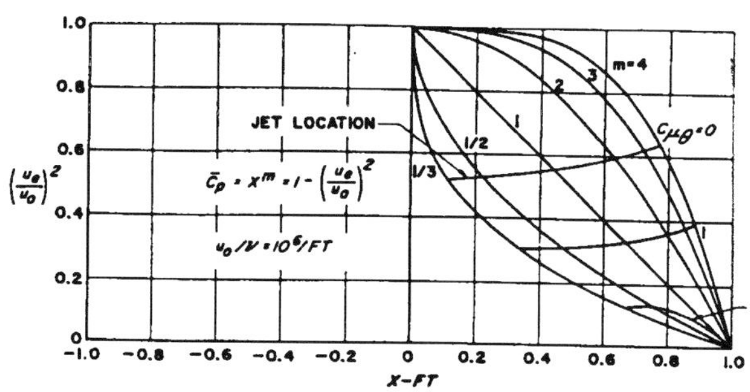

Figure 52 Effect of tangential blowing on location of separation. Blowing slot is located at point where $c_f = 0.001$. Blowing velocity ratio $u_j/u_e = (10)^{1/2}$ Initial flat plate run = 1 ft, turbulent; $u_o/\upsilon = 10^6$/ft, Gartshore-Newman type of calculation.

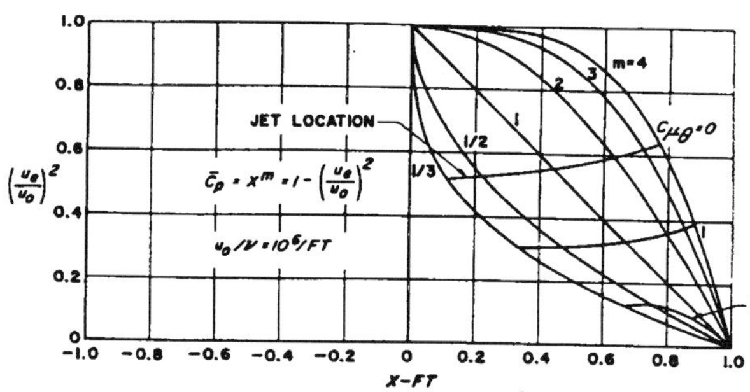

Figure 53 Effect of tangential blowing on location of separation. Blowing slot is located at point where $c_f = 0.001$. Blowing velocity ratio $u_j/u_e = (10)^{1/2}$. Initial flat-plate run = 1 ft, turbulent; $u_o/\upsilon = 10^7$/ft, Gartshore-Newman type of calculation.

jet boundary-layer profile developed a hollow. Hence, the values shown in Figs. 52 and 53 should not be considered much more than estimates, or a start toward more precise analysis. The blowing slot is located at the $C_{\mu\theta} = 0$ line, which is the $c_f = 0.001$ locus. Then, as blowing is increased, separation is delayed to the contours noted. Blowing with an intensity $C_{\mu\theta} = 2.0$ appears to delay separation nearly to the stagnation point. Blowing at the $c_f = 0.001$ line with a $C_{\mu\theta}$ value of 1.0 is just enough to make up for the momentum loss at the slot location. Hence the separation locus should be approximately the same as for the case of natural separation when the initial boundary layer is very thin. If one examines Figs. 19 and 20 for the flat-plate run equal to 1/64 ft, he sees that separation locus is not much different from the $C_{\mu\theta} = 1$ loci in Figs. 52 and 53.

6.4 Inverse Problem for Jet Airfoils

A kind of lifting efficiency can be defined for airfoils. It is illustrated in Fig. 54. The efficiency is simply the ratio of the area contained within the canonical-pressure-distribution loop to that of the circumscribed rectangle. The falloff in pressures toward the rear on the upper surface is necessitated to varying degrees by flow-separation requirements. If the figure represented were a design pressure distribution for some problem, the upper $\bar{C}_p$ bound might be determined by Mach number, for example. Two pressure distributions are shown in Fig. 54, one for a slightly cambered airfoil at angle of attack, the other for a cambered Joukowski airfoil. The cambered airfoil has the higher efficiency. Obtaining lift by angle of attack, according to the definition of efficiency, is inefficient. Because the start of pressure rise can be delayed more if design lifts are low, the efficiency at low lift coefficients is often higher than that of shapes either designed for or operating at high lift.

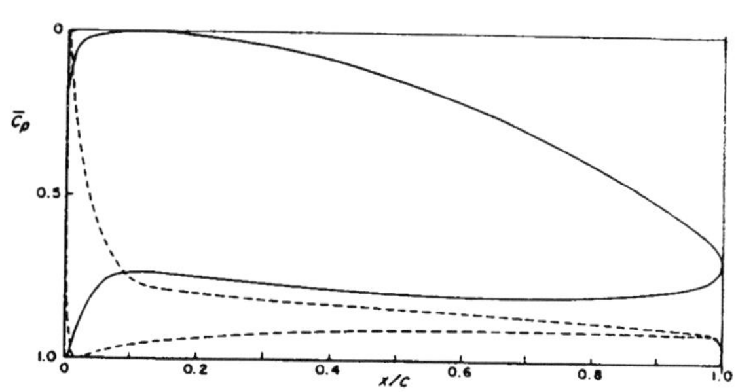

Figure 54 Illustration of the concept of lifting efficiency. Pressure distributions for two airfoils are shown.

Powered lift can improve the efficiency markedly. When there is a curved jet, load can be carried by the jet, and it becomes possible to carry load all the way to the trailing edge. Thus the pressure-recovery problem can be eliminated entirely.

A highly accurate theory of jet-flap airfoils that makes it possible to realize that benefit has been worked out in the past few years by Lopez and his co-workers.[47,48] One of his co-workers, Halsey,[48] reviews the airfoil work in general. But here we feel we have space only to mention one aspect of it, a method that appears to have produced an interesting and significant advance. He has developed an inverse method for jet airfoils. That is, the method specifies the pressure distribution and the jet momentum coefficient and finds the airfoil shape and the jet leaving angle necessary to generate that shape.

The method is not a classical inverse method, such as James has developed for conventional airfoils. Instead, it is a kind of iterative procedure that involves finding a mean line and obtaining modified shapes until convergence is reached. Recourse can be had to the exact direct method for checking purposes. The effects of jet entrainment on the pressure distribution are accounted for approximately. The procedure has been found to produce rapid convergence and accurate results. We can do no better than to quote from Ref. 48 to show the specific steps that are taken in finding a shape. The steps, slightly paraphrased, are as follows:

1) Specify the desired airfoil velocity distribution and jet momentum coefficient.
2) Divide the velocity distribution into symmetric and antisymmetric components.
3) Determine the approximate jet-entrainment effects. Since the operation is noniterative, the sink distribution representing the jet is placed on an extension of the airfoil chord, and its strength is determined by assuming that the velocity just outside the jet is equal to the freestream value. The entrainment effects at the true airfoil surface are assumed to be the same as those calculated at the airfoil chord line.
4) Use an inverse method to find the thickness distribution that corresponds to the symmetrical component of the velocity specified in step 2 that remains after entrainment effects have been subtracted out.
5) Design the mean line needed to produce the antisymmetric portion of the velocity distribution, including the influence of the jet. For that purpose, a linearized method used. This part of the design procedure is the only part that is fundamentally different from conventional design methods. For more details, see Ref. 48. Either: a) the vorticity at the trailing

edge can be specified, find the jet angle; or b) the jet angle can be specified, find the vorticity at the trailing edge.

6) Combine camber and thickness distributions.
7) Using the general nonlinear direct jet-flap airfoil method, check the result. If necessary, repeat the cycle with pressures slightly modified to correct errors revealed by the direct-method check.

We present two figures to show typical results. Figure 55 shows the effect of blowing. High blowing rates relieve the camber, assuming that the velocity distribution is maintained constant. The points are checks of the computed inverse shape by means of the direct, nonlinear method using it as input data. The difference in c_l noted for the three airfoils is due to the lift component of the jet. Each airfoil is shown at the angle required to develop the design pressure distribution.

An extreme case, one that taxes the linearized camber treatment, is shown in Fig. 56. Here entrainment effects have been included in the inverse method. The entrainment effects, as well as the entire camber-shape solution, are worked out along the chord line. The final jet angle is at an inclination of -74.4° to the chord

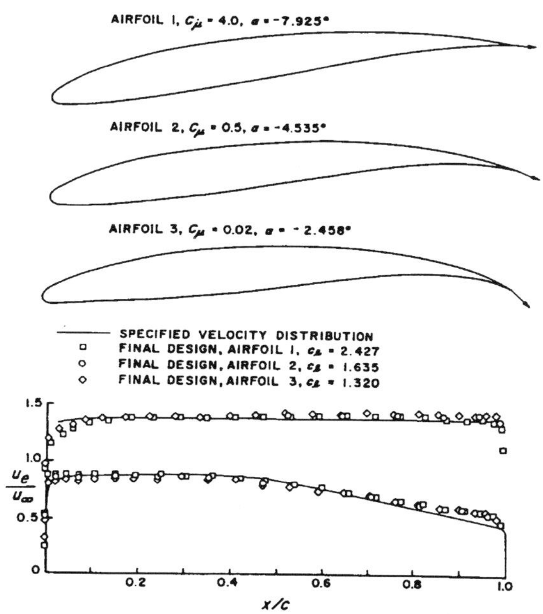

Figure 55 Effect of C_μ on shape of airfoils all designed to have the same upper-surface velocities. Blowing is in direction of mean line at the trailing edge, and indicated by the arrows. Each airfoil is shown at its design angle of attack.

plane, and the airfoil is nose down 4.7° to develop the design pressure distribution. Agreement between the exact check calculation and the linear design method is fair. Considering the extreme nature of this problem, the result is satisfactory.

7. The Future

It was my original intention in planning this lecture to devote considerable attention to the state of knowledge about three-dimensional flows and high-lift flows in the transonic region. However, I shall not, for two reasons. First, I am already exceeding the desired length for this paper, and second, there is not much in the way of theory and analysis methods that can be called solid. Low-speed, two-dimensional airfoil high-lift theory has become a fairly "hard" science, but

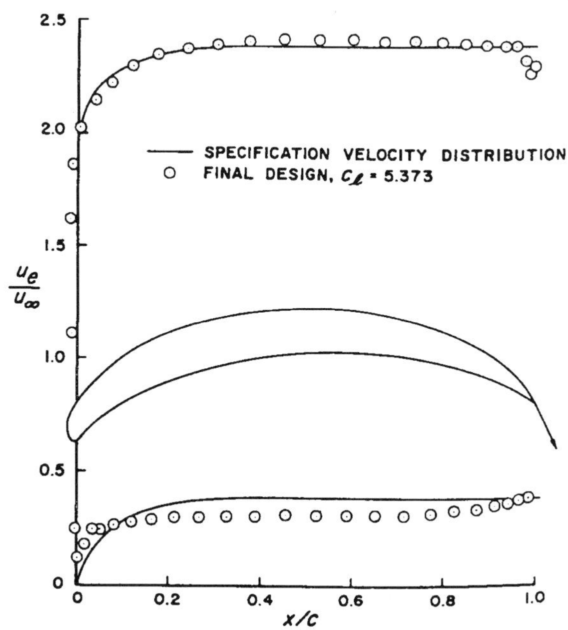

Figure 56 A highly cambered jet-flapped airfoil designed by the inverse method, including entrainment effects. The solid line is the specified velocity distribution. The o points show check results from the exact nonlinear direct method of calculation.

three-dimensional and transonic theory is still no more than a "soft" science, if that much.

It is only necessary to read Callaghan's AGARD lecture notes[19] to realize the diversity of problems that arise as we come closer to real aircraft configurations and to realize the kind of practical empiricism that one is forced to use in applied aircraft design.

Like the two-dimensional problem, the three-dimensional is an analysis involving a combination of inviscid- and boundary-layer flow. The inviscid analysis is nearly here. That is, lifting inviscid-potential-flow solutions can now be calculated readily for as many as 1000 coordinate points. That number will easily solve nearly any simple wing-alone geometry and is rather good at handling wing-fuselage combinations. 1000 points is marginal for a two-element wing alone, but it can produce useful information. To handle three-element wings together with major body effects at least 2000-point solutions are needed. In a private communication, we learn from Norman Donaldson that Bell Aircraft has extended the method to handle 3000 data points. The work is done on a modified UNIVAC 1108 computer. A problem of this size requires more than two hours of machine time. It is hoped that cost per data point continues to decrease.

The boundary-layer solution lags behind the inviscid-flow solution. Although there are formulations and methods for analyzing three-dimensional laminar and turbulent flows—and even a book[49] on the subject—none is powerful enough to analyze the complicated problems of applied analysis that occur, for example, those of swept wings. Progress in fundamental work is encouraging, however. It appears that extensions of the eddy-viscosity formulations used quite successfully for two-dimensional flows do not lose accuracy when extended to three-dimensional flows.[50] In the case of turbulent flow, much empiricism is involved, and the accuracy of a method can be determined only by extensive comparison with experiment. A milestone in the development of the art of computing two-dimensional turbulent boundary layers was the Stanford Conference,[5] in which each participant who claimed to have a good boundary-layer method was asked to calculate a large variety of flows, all from standardized input data under prescribed rules. Sometime in the rather distant future, that kind of conference will be needed to establish the accuracy of methods of calculating three- dimensional flow.

Beyond that problem is the problem of calculating maximum lift, as contrasted with the problem of calculating initial separation. If the calculations of three-dimensional boundary layers are forced to stop when the first separation is encountered, they are likely to do more than barely start. In two dimensions, we can now predict separation reasonably well. With that problem out of the way, the problem of predicting flows with partial separation comes to the front. Ideas being explored range all the way from direct numerical attempts at solving the complete Navier-Stokes equation to approximating the separated region by an inviscid flow.

In the transonic regime, where separation involves shocks, again there is no method of prediction that can be called general. There is a large traffic in papers on shock-boundary-layer interaction, but so far as the author knows, the problem has not been solved in any general sense. A vigorous attack is under way that uses finite-difference methods. Possibly practical methods of analysis will result.

One could talk indefinitely about other special problems related to high lift but that won't do, so I shall end by listing what I think are the ten most important basic theoretical problems of high-lift aerodynamics:

1) Very general calculation of three-dimensional laminar and turbulent flows.
2) Calculation of flows involving partial separation in the rear.
3) Practical calculation of flows involving forward separation bubbles.
4) Practical calculation of flows involving shock-boundary-layer interaction.
5) Calculation of the viscous flow on airfoils and bodies of revolution of length c from about $0.95c$ to $1.05c$. In that region of very rapid change and strong interaction, boundary- layer equations do not apply. With our recently acquired ability to calculate well the forward 95%, our inability to solve the next 10% is a real irritant. Beyond 105% chord, we again can usually do an acceptable job.
6) Further development of inverse methods.
7) Drag of multielement airfoil systems. Drag predictions have a relative error one order of magnitude greater than that of lift predictions. Since propulsive and acceleration information is crucial to aircraft design, that is a very important problem.
8) Practical calculation of merging boundary-layers, wall jets and wakes. Although methods have been developed, it is felt that considerable further development is needed.
9) The analysis of flows over swept wings on which a leading-edge vortex is developed. When the vortex develops, conventional calculations are about 100% in error.

10) Three-dimensional transonic calculations, particularly for arbitrary wings and wing-body combinations.

In several of those problems, the first phase is two-dimensional analysis, but as soon as that is accomplished the three-dimensional looms up as the next target.

As you can see from the list, aerodynamics still has a long way to go before we can truly calculate, instead of just estimate. Yet in retrospect, we see that we have progressed a long way toward the goal in the past thirty years. Perhaps another Wright Brothers lecture should be given on this same subject 20 to 30 years from now. With the computer revolution that we have, the prospects of finding practical calculation methods for most of the ten problems are bright. We hope there will be some new "inventions" in flow and flow control. Cornish's spanwise blowing[52] might be one, and effective application of the trapped-vortex idea[53] might be another.

In October 1908, at a banquet in his honor in France and after very successful flights in September, Wilbur Wright was unexpectedly called on to say a few words. He said, "I know of only one bird, the parrot, that talks, and it can't fly very high." I, too, cannot fly very high and I have talked too much.

References

[1]Prandtl, L., "Über Flüssigkeitsbewegung bei sehr kleiner Reibung," III. Internationaler Mathematiker-Kongreß, Heidelberg, 1904.

[2]Weyl, A. R., "High-Lift Devices and Tailless Airplanes," *Aircraft Engineering*, Oct.1945, p.292.

[3]Hagerdorn, H. and Ruden, P., "Wind Tunnel Investigations of a Wing with Junkers Slotted Flap and the Effect of Blowing Through the Trailing Edge of the Main Surface over the Flap," RAE Translation No.442, Dec.1953, Royal Aircraft Establishment, Farnborough, England.

[4]"Experiments on Models of Aeroplane Wings at the National Physical Laboratory," R&M No. 110, March 1914, Aeronautical Research Council, London.

[5]Page, F. H., "The Handley Page Wing," *The Aeronautical Journal*, June 1921, p.263.

[6]Glauert, H., "The Handley Page Slotted Wing," R&M No.834, Mar.1922, Aeronautical Research Council, London.

[7]Lachmann, G. V., *Boundary Layer and Flow Control*, Vol.1, Pergamon Press, London, 1961.

[8]LePage, W. L., "Further Experiments on Tandem Aerofoils," R&M No.886, May 1923, Aeronautical Research Council, London.

[9]Harris, R. G. and Bradfield, F. B., "Model Experiments with Variable Camber Wings," R&M No.677, June 1920, Aeronautical Research Council, London.

[10]Fowler, H. D., "Variable Lift," *Western Flying*, Nov. 1931, p. 31.

[11]Weick, F. E. and Platt, R. C., "Wind-Tunnel Tests of the Fowler Variable-Area Wing," Tech. Note 419, May 1932, NACA.

[12]Alston, R. P., "Wing Flaps and Other Devices as Aids to Landing," *Journal of the*

Royal Aeronautical Society, Vol. 39, 1935, p.637.

[13]Young, A. D., "The Aerodynamic Characteristics of Flaps," R&M No.2622, 1953, Aeronautical Research Council, London.

[14]Duddy, R. R., "High-Lift Devices and Their Uses," *Journal of Royal Aeronautical Society*, Vol.53, 1949, p.859.

[15]Cleveland, F. A., "Size Effects in Conventional Aircraft Design," *Journal of Aircraft*, Vol.7, Nov.-Dec. 1970, pp.483-512

[16]Faulkner, S. M., Hess, J. L., Smith, A. M. 0., and Liebeck, R. H., "Charts and Formulas for Estimating Velocity Fields in Incompressible Flow," Rept. LB 32707, July 1968 (AD 772 233), McDonnell Douglas Corp., Long Beach, Calif.

[17]Mayer, J. P., "A Limit Pressure Coefficient and an Estimation of Limit Forces on Airfoils at Supersonic Speeds," Research Memo. L8F23, 1948, NACA.

[18]Smith, A. M. 0., "Aerodynamic Lifting Potentialities of Aircraft in Subsonic Flight, Rept. ES 20930, 1947, Douglas Aircraft Co., Long Beach, Calif.

[19]Callaghan, J. G., "Aerodynamic Prediction Methods for Aircraft at Low Speeds with Mechanical High Lift Devices," AGARD Lecture Series No. 67, May 1974, von Karman Institute, Brussels, Belgium.

[20]Smith, A. M. 0., "Aerodynamics of High-Lift Airfoil Systems," *Fluid Dynamics of Aircraft Stalling*, AGARD-CP-102, 1972.

[21]Assessment of Lift Augmentation Devices," AGARD Lecture Series No.43, AGARD LS-43-71 (Available as AD 720-259, 1971).

[22]Thain, J. A., "Reynolds Number Effects at Low Speeds on the Maximum Lift of Two-Dimensional Aerofoil Sections Equipped with Mechanical High-Lift Devices," *Quarterly Bulletin National Aeronautical Establishment*, Ottawa, Rept. DM/NAE, 1973.

[23]Cebeci, T., Mosinskis, 0. J., and Smith A. M. 0, "Calculation of Separation Points in Incompressible Turbulent Flows," *Journal of Aircraft*, Vol.9, Sept, 1972, p.618-624.

[24]Loftin, L. K., Jr. and von Doenhoff, A. E., "Exploratory Investigation at High and Low Subsonic Mach Numbers of Two Experimental 6-Percent Thick Airfoil Sections Designed to Have High Maximum Lift Coefficients," RM L51F06, 1951, NACA.

[25]Stratford, B. S., "The Prediction of Separation of the Turbulent Boundary Layer," *Journal of Fluid Mechanics*, Vol.5, 1959, pp.1-16.

[26]Schlichting, H., *Boundary-Layer Theory*, 6th ed. McGraw-Hill, New York, 1968.

[27]Liebeck, R. H., "A Class of Airfoils Designed for High Lift in Incompressible Flow," *Journal of Aircraft*, Vol. 10, Oct. 1973, p. 610-617.

[28]James, R. M., "A New Look at Two-Dimensional Incompressible Airfoil Theory," Rept. MDC J0918/01, May 1971, Douglas Aircraft Co., Long Beach, Calif.

[29]Gartshore, I. S., "Predictions of the Blowing Required to Suppress Separation from High Lift Airfoils," *CASI Transations*, Vol.4, March 1971, p.39.

[30]Stevens, W. A., Goradia, S. H., and Braden, J. A. "Mathematical Model for Two-Dimensional Multi-Component Air foils in Viscous Flow," NASA Contractor Rept. CR-1843, July 1971, NASA.

[31]Stevens, W. A., Goradia, S. H., and Braden, J A. "Mathematical Model for Two-Dimensional Multi-Component Airfoils in Viscous Flow," AIAA Paper 72-2, San Diego, Calif., 1972.

[32]Foster, D. N., Ashill, P. R., and Williams, B. R., "The Nature Development and Effect of the Viscous Flow Around an Airfoil with High-Lift Devices," Tech. Rept. 72227, Jan.

1973, Royal Aircraft Establishment, Famborough, England.

[33]Prandtl, L. and Tietiens, 0. G., *Applied Hydro- and Aeromechanics*, Dover, New York, p.155.

[34]Abbott, I. H., and von Doenhoff, A. E., *Theory of Wing Sections*, Dover, New York, 1958, p.227.

[35]Perkins, C. D. and Hage, R. E., *Airplane Performance, Stability, and Control*, Wiley, New York, 1954, p.78.

[36]Chen, A. W., "The Determination of the Geometries of Multiple-Element Airfoils Optimized for Maximum Lift Coefficient,' TM-X-67591, 1971, NASA.

[37]Liebeck, R. H. and Smyth, D. N., "Study of Slat-Airfoil Combinations Using Computer Graphics," *Journal of Aircraft*, Vol.10 April 1973, p.254-256.

[38]Hess, J. L. and Smith, A. M. 0., "Calculation of Potential Flow about Arbitrary Bodies," *Progress in Aeronautical Science*, Vol.8 Pergamon Press, New York, 1966.

[39]Callaghan, J. G. and Beatty, T. D., "A Theoretical Method for the Analysis and Design of Multielement Airfoils," *Journal of Aircraft*, Vol.9, Dec.1972, p.844-848.

[40]Gentry, A. E., "How Sails Work," *SAIL Magazine* (in several issues between April 1973 and Oct.1973).

[41]Ashill, P. R., "A Study of the Effect of the Wake of the Main Aerofoil of a Fowler Flap Configuration on the Lift of the Flap," Tech. Rept. TR 72081, July 1972, Royal Aircraft Establishment Farnborough, England.

[42]Moser, A. and Shollenberger, C. A., "Inviscid Wake-Airfoil Interaction on Multielement High Lift Systems," *Journal of Aircraft* Vol.10, Dec.1972, p.765-770.

[43]Wilkinson, D. H., "A Numerical Solution of the Analysis and Design Problems for the Flow Past One or More Aerofoils or Cascades," R&M 3545, 1967, Aeronautical Research Council London.

[44]Cebeci, T. and Smith, A. M. 0., Analysis of Turbulent Boundary Layers, Academic Press, New York, 1974.

[45]Cebeci, T. and Witherspoon, G. F., "Theoretical Suction and Pressure Distribution Bounds for Flow Separation in Retarded Flow," *Journal of Aircraft*, Vol.11, Jan.1974, pp.61 -64.

[46]Gartshore, I.S. and Newman, B. G., "The Turbulent Wall Jet in an Arbitrary Pressure Gradient," *The Aeronautical Quarterly*, Feb 1969, p.25.

[47]Shen, C. C., Lopez, M. L., and Wasson, N. F., "A Jet-Wing Lifting-Surface Theory Using Elementary Vortex Distributions," *Journal of Aircraft*, Vol.12, May 1975, pp.448-456.

[45]Halsey, N. D., "Methods for the Design and Analysis of Jet Flapped Airfoils," *Journal of Aircraft*, Sept.1974, pp. 540-546.

[49]Nash, J. F. and Patel, V. C., *Three-Dimensional Turbulent Boundary Layers*, S. B. C. Tech. Books, Atlanta, Ga., 1972.

[55]Cebeci, T., "A General Method for Calculating Three-Dimensional Incompressible Laminar and Turbulent Boundary. Layers; II. Three-Dimensional Flows in Cartesian Coordinates." Rept. MDC J6517, Mar. 1974, Douglas Aircraft Co., Long Beach. Calif.

[51]Kline, S. J., Morkovin, M. V., Sovran, G., and Cockrell, D. S., "Computation of Turbulent Boundary Layers," AFOSR-IFFS-Stanford Conference, Stanford Univ. Press, 1968.

[52]Cornish, J. J., III, "High-Lift Applications of Spanwise Blowing," Paper No.70-09, *Seventh Congress of the ICAS*, Rome 1970.

[53]Krall, E. M., and Haight, C. H., "Wind Tunnel Tests of a Trapped-Vortex High-Lift Airfoil," Rept. B-94300/3TR- 10, Advanced Technology Center, Dallas, Texas.

1950 Lovell Lawrence Jr.
1951 Robert C. Truax
1952 Richard W. Porter
1953 David A. Young
1954 A. M. O. Smith
1955 E. N. Hall
1956 Chandler C. Ross
1957 Thomas F. Dixon
1958 Richard B. Canright
1959 Samuel K. Hoffman
1960 von Kármán
1961 von Braun
1962 Robert R. Gilruth
1927 Col. Charles A. Lindbergh, USAAC
Charles Matthews Manly
(posthumously)
R. Adm. Richard Evelyn Byrd, USN
Dr. Joseph Sweetman Ames
Dr. Jerome C. Hunsaker
1960 Dr. Robert H. Goddard
(posthumously)
1962 Dr. Hugh Latimer Dryden
1964
ROBERT H. GODDARD MEMORIAL AWARD
Established in 1947, the Goddard Award con-

AERONAUTICS AND
INSTITUTE

7

High-Lift Airfoils

by

Robert H. Liebeck
A member of AMO's Research Group

After receiving a bachelor's degree in aeronautical engineering from the University of Illinois in 1961, I began work in the Missile and Space Systems Division of the Douglas Aircraft Company in Santa Monica for the summer. In the fall, I returned to Illinois to work on a masters degree, and the process of working summers at Douglas and spending the school year at Illinois continued for my entire career as a graduate student. Allen Ormsbee, my advisor at Illinois, suggested a thesis topic of airfoil design for minimum drag which I began work on in 1965. Selection of this topic led to my transfer to AMO's Aerodynamic Research Group in the Douglas Aircraft Division in Long Beach in 1967. When I arrived in Long Beach in June 1967 my thesis was in trouble. While I had in principle formulated the minimum drag problem, a unique solution was not apparent. AMO said, "Why don't you see how much lift you can get from an airfoil?" Peter Lissaman, a Caltech professor and consultant to AMO's group provided valuable guidance. By the end of that summer, the high-lift problem was formulated. I returned to Illinois in September, and coupled with the airfoil design work under Ormsbee, I was able to complete the thesis *Optimization of Airfoils for Maximum Lift* by Christmas. Having, AMO, Al, and Peter as advisors and mentors was priceless.

Returning to Long Beach in the spring of 1968, AMO provided the luxury and opportunity for me to continue work on the high-lift problem as my primary assignment. This was ideal environment for such work. John Hess had the panel method in good working order, and Tuncer Cebeci and AMO had just demonstrated remarkable success in predicting airfoil drag with their finite-difference + eddy viscosity boundary-layer method.

During this same period, Malcolm James was developing a solution technique for the airfoil problem in potential flows. Completed in 1970, the James method produced an exact inverse solution for the airfoil design problem. For the high-lift airfoil problem, I had been using a second-order linear theory to obtain the airfoil geometry corresponding to a maximum lift velocity distribution. The resulting airfoils were highly cambered and clearly beyond the level of approximation of linear theory. Malcolm's inverse technique provided the exact geometry, and this allowed the high-lift airfoils to become real.

An array of high-lift designs were developed in 1971, and in early 1972. The first wind tunnel test was conducted at McAir in St. Louis. The test airfoil achieved a $C_{l_{max}}$ of 2.18 and a maximum LID of 235 at a Reynolds number of 106, which was believed to be a record at the time. Comparison of the wind tunnel results with the theoretical predictions of the panel method coupled with the boundary layer method showed excellent agreement.

Thus, with the combination of the James method for airfoil design, the Hess-Smith panel method for potential flow analysis, and the Cebeci-Smith boundary-layer method, a complete system for airfoil design and analysis was available - totally within AMO's Aerodynamic Research Group. This capability, together with direct access to those who created it, was fundamental to the success of my high-lift design work.

Beyond the high-lift airfoil work, it should be noted that in 1968, virtually all of the aerodynamic design and analysis methods in use at the Douglas Aircraft Company were -spawned in AM Aerodynamic Research Group. AMO himself was a fundamental contributor to, and in most cases the originator of said methods. AMO retired in 1975, and this began the transition to a dependence on NASA and a few select university professors for aerodynamic methodology. These sources have provided some excellent capability, however, the Company awaits the arrival of the next AMO.

A CLASS OF AIRFOILS DESIGNED FOR HIGH LIFT WITHOUT SEPARATION IN INCOMPRESSIBLE FLOW*

by

R. H.. Liebeck and A.M.O. Smith

Douglas Aircraft Company

Abstract

The problem studied is that of designing a single element airfoil which provides the maximum possible lift in an unseparated incompressible flow. First, an airfoil velocity is defined and optimized using boundary-layer theory and the calculus of variations. The resulting velocity distribution is then used as an input for an inverse airfoil design program which provides the corresponding airfoil shape. Since there is no guarantee that an arbitrarily defined velocity distribution will yield a physically possible airfoil shape, some parametric adjustments in the optimized distributions are required in order to obtain realistic and practical airfoil geometries. A typical result is a profile designed for a freestream Reynolds number of five million which has a lift coefficient of 2.2 and a corresponding drag coefficient of 0.006. Although compressibility effects have not been considered directly, the character of the resulting velocity distributions is such that the critical Mach number is in effect maximized.

1. Introduction

The basic problem considered in this study is that of obtaining an answer to the question: "What is the maximum lift coefficient which can be obtained from a monoelement airfoil, and what is the shape of that airfoil?" This question is clearly quite general, and it is necessary to qualify it to an extent that an engineering problem can be formulated and solved. To begin with, most airfoils are partially separated when they reach their absolute maximum lift coefficient, and their continued operation under such conditions is usually con-

*Douglas Aircraft Company Report, MDC J-1097/01, Dec. 1971

sidered impractical. Moreover, the development of techniques for the analysis of separated flow has not reached a level of precision and reliability where meaningful and realistic optimization techniques can be applied. Thus, it seems reasonable, at least initially, to consider the problem of the design of a monoelement airfoil which provides the maximum lift in an unseparated flow.

Another important consideration in the design of a high-lift airfoil is the effect of compressibility. It is well known that even though an airfoil may be operating at a low flight speed, the super velocities on its upper surface can be quite high—particularly at high lift coefficients. In some cases the maximum lift coefficient for conventional airfoils is limited by compressibility effects even for flight Mach numbers on the order of 0.15. However, for the purposes of the present analysis, the flow is initially assumed to be incompressible, and compressibility effects will be discussed later.

The basic approach for the design of a maximum lift monoelement airfoil will be to first specify a velocity distribution and then calculate the corresponding airfoil shape. The velocity distribution must satisfy the three basic criteria:

1. The boundary layer does not separate.
2. The corresponding airfoil shape is practical and realistic.
3. Maximum possible C_L is obtained.

This may be interpreted as a form of variational problem where an extremum of C_L is sought subject to the constraints imposed by 1 and 2 above. In this study such a variational problem is formulated and solved using existing boundary-layer and potential-flow analyses as described in Ref. 1.

2. Results

The analysis of Straftord provides a pressure distribution which recovers a given pressure difference in the shortest possible distance without separation, and therefore for the purpose of maximizing the lift, it is logical to employ such a distribution for pressure recovery on the upper surface of the airfoil. Two basic cases are considered:

Case (a): A turbulent boundary layer exists over the entire upper surface of the airfoil.

Case (b): A laminar boundary layer exists from the leading stagnation point s_p to the maximium velocity point s_o with instantaneous transition to a turbulent boundary layer at s_o.

The basic requirements on the airfoil geometry are that it is non-reentrant and that it has a rounded leading edge and sharp trailing edge. According to the variational analysis, for maximum lift it is desirable to obtain an airfoil velocity distribution which is as close as possible to that indicated by the solid line in Fig. 1. By strictly following this approach, the airfoil shape

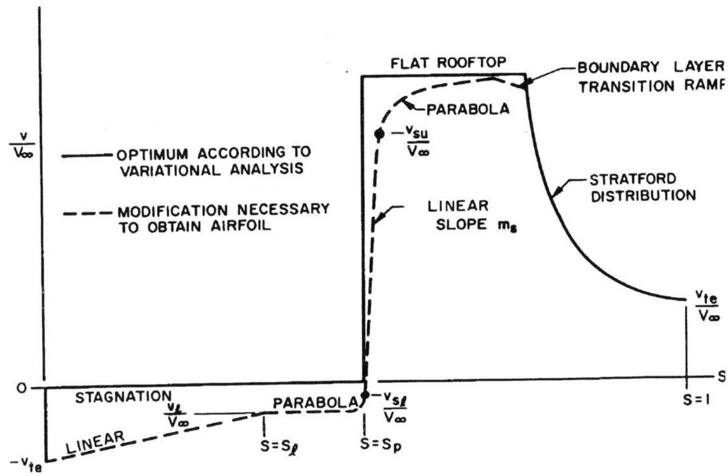

Figure 1. Optimum form for airfoil velocity distributin including the modification necessary for obtaining an airfoil shape.

and corresponding velocity distribution of Fig. 2 was obtained for case (b) with a freestream Reynolds number of five million. This result is quite significant from a purely theoretical point of view: it implies that the maximum lift coefficient which can be obtained from a monoelement airfoil in an unseparated incompressible flow at a Reynolds number of five million is about 3.0.

From a practical standpoint, the airfoil of Fig. 2 does not appear to be very useful. Its sharp leading edge will tend to cause separation when the airfoil is operated at angles of attack other than the design value, and it is unacceptably thin as far as structural considerations are concerned.

Consequently, a set of sample airfoil designs have been computed where a reduction in the value of the lift coefficient has been accepted in order to obtain reasonable performance over an angle of attack range together with an airfoil shape which is structurally feasible. For case (a), where the bound-

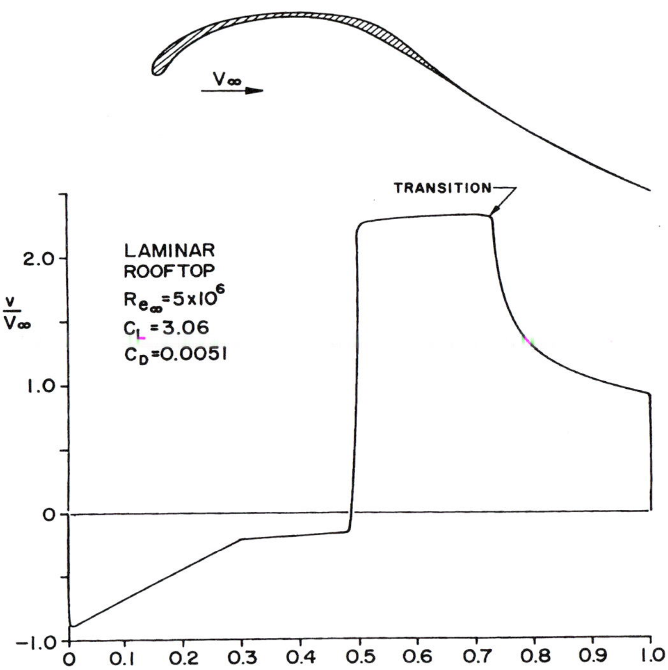

Figure 2. Velocity distribution and corresponding airfoil shape designed neglecting practical considerations. Case (b), $Re_\infty = 5 \times 10^6$. See Figure 7 for practical result.

ary layer is taken as turbulent over the entire airfoil surface, designs for freestream Reynolds numbers of 5, 10, and 20 million are shown in Figs. 3, 4, and 5, respectively, and similarly for case (b), where the boundary layer is assumed laminar over the entire lower surface and accelerating portion of the upper surface with transition to a turbulent boundary layer occurring along the ramp at the upper surface pressure peak, designs for Reynolds numbers of 1, 5, and 10 million are shown in Figs. 6, 7 and 8, respectively. At Reynolds numbers lower than 5 million it is doubtful that fully turbulent flow could be achieved for case (a), and above 10 million early transition will probably occur for case (b).

Reference 1 contains tables of coordinates for the airfoils of Figs. 3 through 8 where the airfoil chord line lies along the x-axis from -1.0 to 0 with the trailing edge at the origin. The design Reynolds number of an airfoil should be regarded as the lowest Reynolds number at which the airfoil can be operated and still have the flow remain attached over the entire surface. For the

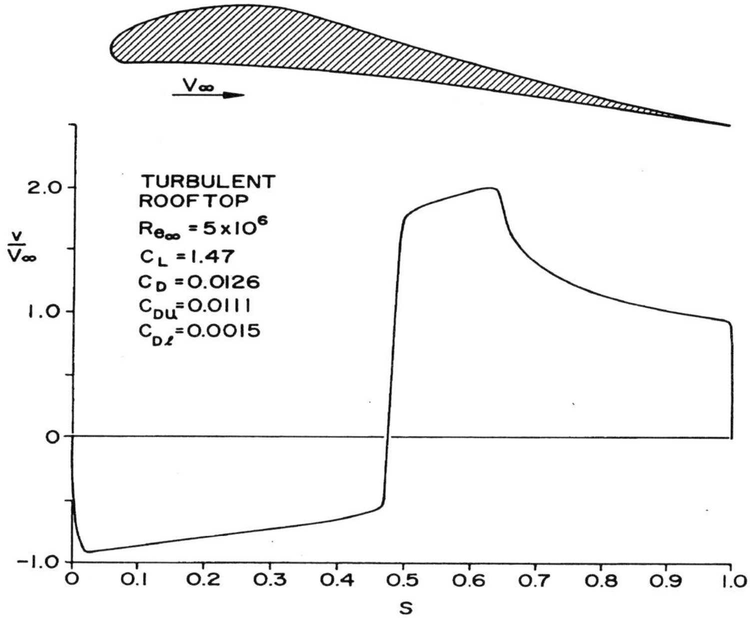

Figure 3. Velocity distribution and corresponding airfoil shape. Case (a), $Re_\infty = 5 \times 10^6$.

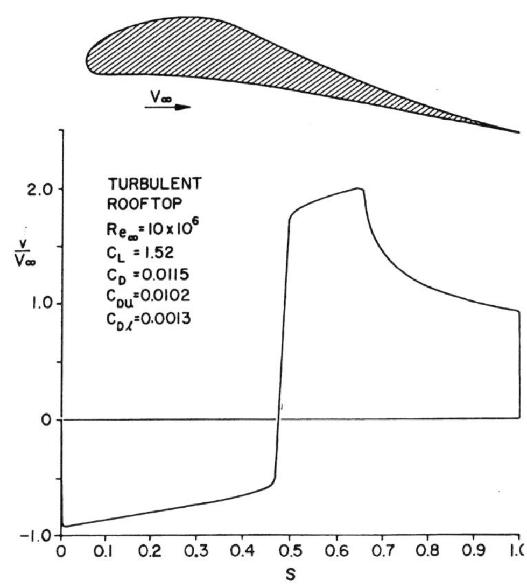

Figure 4. Velocity distribution and corresponding airfoil shape. Case (a), $Re_\infty = 10 \times 10^6$.

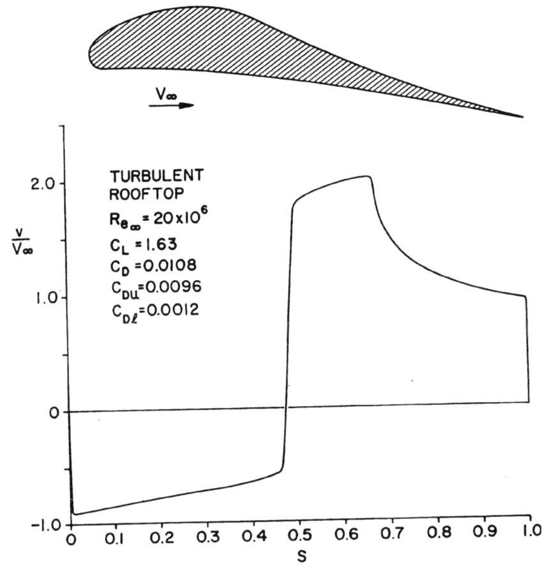

Figure 5. Velocity distribution and corresponding airfoil shape. Case (a), $Re_\infty = 20 \times 10^6$.

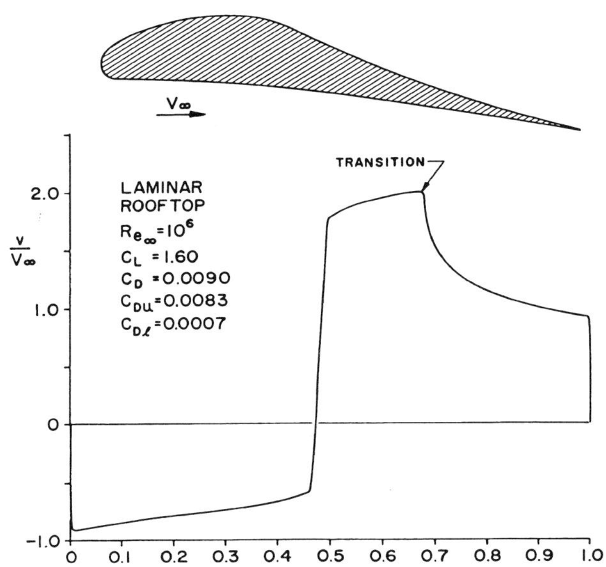

Figure 6. Velocity distribution and corresponding airfoil shape. Case (b), $Re_\infty = 10^6$.

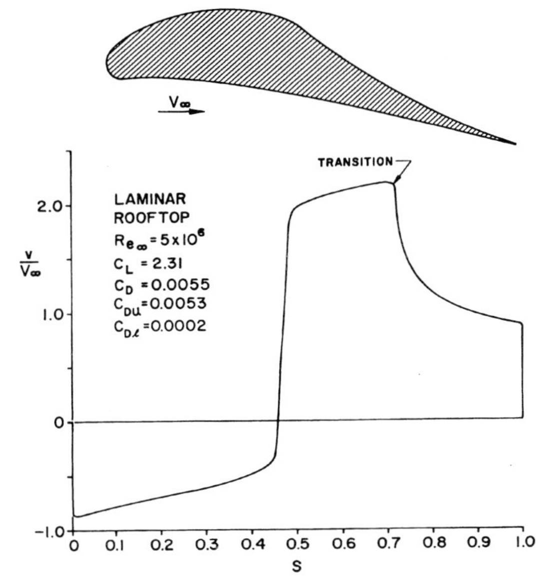

Figure 7. Velocity distribution and corresponding airfoil shape. Case (b), $Re_{\infty} = 5 \times 10^6$.

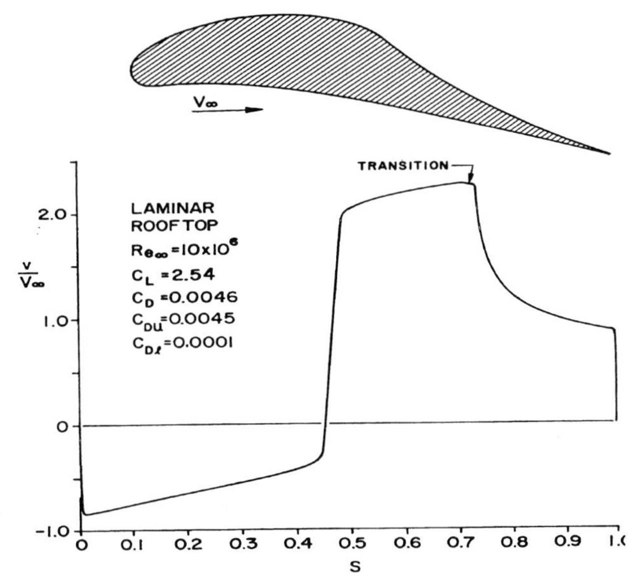

Figure 8. Velocity distribution and corresponding airfoil shape. Case (b), $Re_{\infty} = 10 \times 10^6$.

airfoils of case (b), however, if the Reynolds number becomes too high, premature transition may occur which would no doubt result in separation.

The boundary-layer behavior of the airfoil of Fig. 7 (laminar rooftop, $Re_\infty = 5 \times 10^6$), has been checked using three independent calculation methods: the Douglas turbulent boundary-layer program[2] (a finite-difference method) the NASA-Lewis boundary-layer program[3] (an integral method), and the Truckenbrodt integral method[4]. All three indicated that the boundary layer does not separate and gave values for the drag coefficient which were virtually the same. The calculated boundary-layer displacement thickness was added to the airfoil shape and the potential flow was then recomputed and the result is shown in Fig. 9. For the remaining airfoils, the drag was calculated using the Truckenbrodt method. A laminar boundary layer was assumed for the entire lower surface for the airfoils of case (b), and a turbulent one for the airfoils of case (a).

The airfoil of Fig. 7 was also rotated through a range of angles of attack

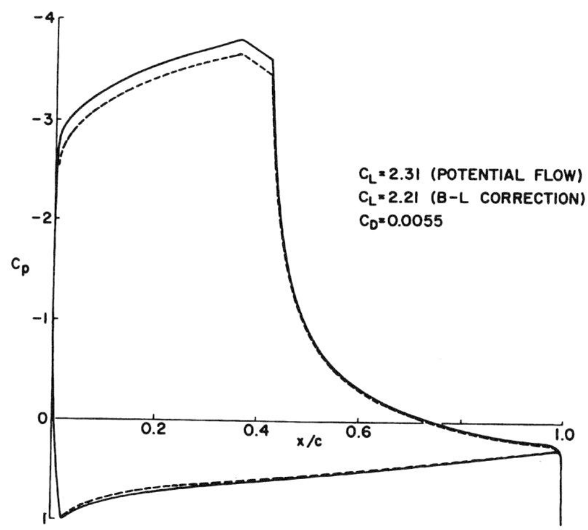

Figure 9. Boundary-layer tyhickness correction for airfoil of figure 7. Case (b), $Re_\infty = 5 \times 10^6$.

to examine its off-design performance. Fig.10 shows the resulting pressure distributions for six angles of attack including the design value. These results would appear to indicate a low drag bucket from an attack of attack of three or four degrees below the design value ($C_L \approx 2.0$) to two degrees above the design value ($C_L \approx 2.5$). At the design angle of attack, the pitching moment about the 1/4-chord is $C_{mc/4} = -0.045$.

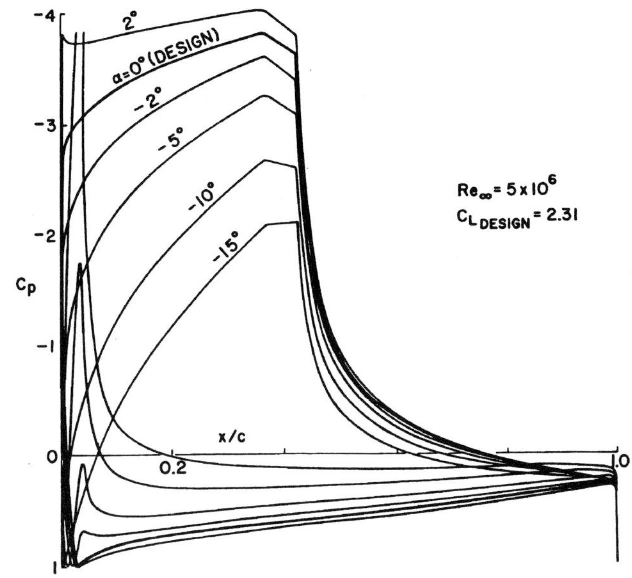

Figure 10. Off-design performance for airfoil of figure 7. Case (b), Re_∞ = 5 x 10⁶.

The following table lists some of the significant aerodynamic parameters for the airfoils of Figs. 3 through 8. $C_L^{3/2}/C_D$ is a so-called "power parameter" for an airplane and the power required for flight decreases as the value of this parameter is increased. (It should be noted that this parameter typically uses the values of C_L and C_D for the entire airplane, while airfoil section data is quoted here.)

Airfoil	C_L	C_D	C_{Du}	C_{Dl}	C_L/C_D	$C_L^{3/2}/C_D$	Maximum Thickness
Case (a)							
Re_{so} = 5 x 10⁶	1.47	0.0126	0.0111	0.0015	117	141	12%
Re_{so} = 10 x 10⁶	1.52	0.0115	0.0102	0.0013	132	163	13%
Re_{so} = 20 x 10⁶	1.63	0.0108	0.0096	0.0012	151	193	14%
Case (b)							
Re_{so} = 1 x 10⁶	1.60	0.0090	0.0083	0.0007	176	222	15%
Re_{so} = 5 x 106	2.31	0.0055	0.0053	0.0002	420	640	18%
Re_{so} = 10 x 10⁶	2.54	0.0046	0.0045	0.0001	462	880	19%

While any direct comparison with a convention al airfoil would be inappropriate, it is noted that NACA 64_3-618 has a maximum C_L/C_D of about 110 and $C_L^{3/2}/C_D = 75$ at $C_L = 0.5$.

4. Conclusions

The most fundamental result obtained from this analysis appears to be that existing aerodynamic theory may be used to obtain airfoil sections which have been optimized for high lift at very low drag. Traditionally, most airfoil design work has been carried out using the direct approach, namely starting with a given airfoil geometry, modify it until the desired performance is obtained. (This is not to say that no airfoil design work has been done using inverse methods. The studies of Wortmann[5] and [6], Eppler[7], and Nonweiler[8] are excellent examples of the application of inverse techniques to airfoil design.) However, it is felt that the present work's significance (including the preliminary studies in Ref. 9) lies in the setting and solution of a variational problem in optimal incompressible airfoil design. That such a problem can be solved in a relatively exact sense, reflects favorably on the current state of boundary-layer and potential-flow theory.

More specifically, the airfoil designs obtained in this study are exact in that they correspond precisely to their potential flow velocity distributions, and, in addition, the boundary-layer analysis has been applied along the airfoil surface itself rather than using the distance along the chord as an independent variable. Therefore, the resulting solutions may be regarded as being as analytically precise as possible using state-of-the-art boundary-layer and potential -flow theory.

Although the airfoil solutions are exact, they are by no means unique as is clearly demonstrated by the airfoils of Figs. 2 and 7. Basically, changes in the form of the leading-edge acceleration velocity distribution affect the corresponding leading-edge radius, and changes in the lower surface distribution influence the resulting airfoil thickness distribution. The example airfoils given in this report have been designed to have reasonable thickness distribution and leading-edge geometries, and it has been required that the flow on the lower surface be continuously accelerating.

The off-design performance shown in Fig. 10 of the airfoil of Fig. 7 indicates that pressure spikes appear on the upper surface when the angle of attack is reduced below the design value. These spikes will tend to cause premature transition of the boundary layer and therefore are regarded as undesirable for the airfoils of case (a). At very low angles of attack (on the order of 10 degrees below the design angle) the lower surface pressure spike may cause local separation for the airfoils of both cases (a) and (b). It is very difficult, if not impossible to prescribe a nose region velocity distribution at a single angle of attack which will not produce a pressure spike when the resulting airfoil is oriented at an angle of attack significantly different from the design value. The best way around this problem is to employ a multiple angle of attack design procedure which is discussed briefly in the next section. Nevertheless, the results of Fig.10 suggest a low drag bucket C_L range from approximately C_L = 2.0 to C_L = 2.5 which is quite reasonable for a high performance airfoil.

The resulting airfoil geometries of case (a) show relatively little dependence on the design Reynolds number as compared with those of case (b) where the geometries are quite different. This may be hueristically explained by the fact that a laminar boundary layer's growth is far more sensitive to the Reynolds number than is the turbulent boundary layer's growth.

Probably the most significant performance feature of the airfoils is that they obtain high lift with very low drag—and not in the value of their lift coefficients alone. There are, of course, many existing airfoils whose values of C_{Lmax} equal or exceed the resulting design lift coefficients of the high-lift airfoils developed in this report. However, the performance of such airfoils is also characterized by relatively high drag at these lift coefficients—on the order of three to five times that of the airfoils developed in this study. This is due to the fact that most conventional airfoils are partially separated when they are operated near or at their C_{Lmax} while the present airfoils are specifically designed to maintain unseparated flow over their entire surface.

Compressibility has not been considered in the foregoing analysis, however, some important observations can be made at this time. The optimized upper surface velocity distributions may also be characterized as having the minimum possible peak velocity for the load ($\int v/V_\infty ds$) which they carry. (It should be noted that this is true for all members of the family of the flat rooftop distributions as shown in Fig. 11, and not just that particular one which maximizes $\overline{v}_t/V_\infty$). Consequently, the airfoils developed in this study may also be regarded as having the highest possible critical Mach number for their design lift coefficients.

Boundary-layer thickness effects do not appear to substantially reduce the performance of the airfoil sections. The results shown in Fig. 9 indicate that the main effect is to reduce the level of the rooftop portion of the pres-

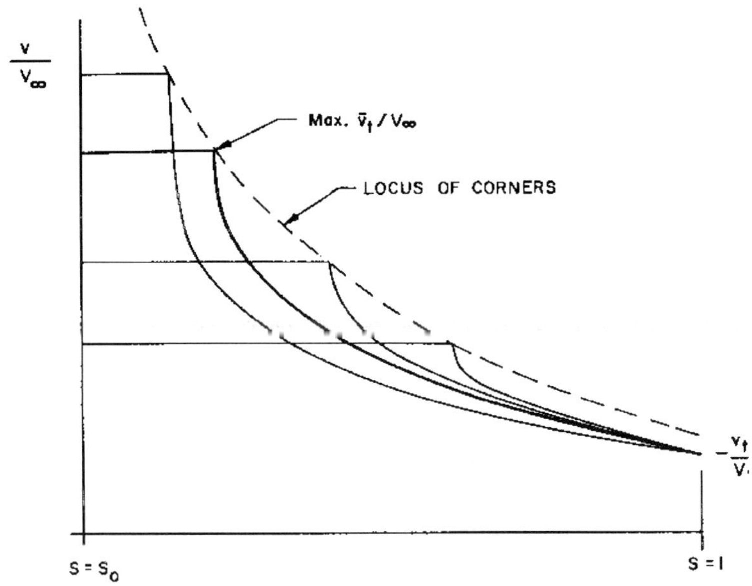

Figure 11. Family of nonseparating flat rooftop velocity distributions, including that particular one which maximizes $\bar{v}_t/v_\infty$. $Re_{\sigma o}$ and v_{te} fixed (not to scale).

sure distribution which could be corrected for by appropriately increasing the rooftop level in the input velocity distribution for the inverse airfoil design program. However, the reduced level will serve to enhance the capability of the boundary layer to survive the Stratford recovery region.

5. Recommendations

The design and analysis of the high-lift airfoils of this study has been carried about as far as possible using some of the latest and most accurate theoretical methods. Consequently, the next logical step is a thorough experimental evaluation through a wind tunnel test program. Assuming that the theoretically predicted performance is verified, some extensions of the present work together with the study of more general airfoil design problems appear to be in order.

To begin, the high-lift airfoils of this study should be refined for off-design performance using a multiple angle of attack design method. Basically, this would involve the specification of the upper surface velocity distribution at the highest α for which low drag performance is desired, and similarly the specification of the lower surface velocity distribution at the lowest α for which low drag performance is desired. By employing the proper transformation, both of these velocity distributions can be related to a single velocity distribution for the airfoil at zero lift which would then be used to calculate the corresponding airfoil shape. An obvious generalization of this approach is the design of minimum drag airfoils over a specified C_L-range. Flat rooftop distributions of the form shown in Fig. 11 would be used for both the upper and lower surfaces with the upper surface distribution being defined for the flight Reynolds number to be encountered when the airfoil is at the high end of its C_L-range, and similarly the lower surface distribution being defined for the Reynolds number of the low end of the C_L-range. It would appear that a thorough solution of this design problem will be a rather extensive project.

The implication that the flat rooftop distributions developed in the present work will have high critical Mach numbers strongly suggests the investigation of the design of subcritical compressible airfoils using this form of distribution. This will require the use of a compressible inverse design method (some of which exist at this time) together with the appropriate modifications to the boundary-layer calculations.

5. References

1. Liebeck, R.H., and Smith, A.M.O.: A Class of Airfoils Designed for High Lift Without Separation in Incompressible FlowDouglas Aircraft Company Report MDC J-10097/01, Dec. 1971.

2. Cebeci, T., Smith, A.M.O., and Wang, L.C.: A Finite-Difference Method for Calculating Compressible Laminar and Turbulent Boundary Layers. Douglas Aircraft Company Report DAC 67131, Pt. I, March 1969.

3. McNally, W.D.: Fortran Program for Calculating Compressible Laminar and Turbulent Boundary Layers in Arbitrary Pressure Gradients. NASA TN D-5681, May 1970.

4. Truckenbrodt, E.: Ein Quadraturverfahren zur Berechnung der laminaren und turbulenten Reibungsschicht bei ebener und rotations-symmetrischer Strömung. Ingen.-Arch., Vol. 20, p. 211, 1952.

5. Wortmann, F. X.: Ein Beitrag zum Entwurf von laminaren Profilen für Segelflugzeuge und Hubschrauber. Z. Flugw., Vol. 3 (10), 1955, Germany.

6. Wortmann, F.X.: On the Optimization of Flap Profiles. Presented at the Eleventh OSTIV Congress in Leszno, Poland, June 1968.

7. Eppler, R.: Laminarprofile für Reynolds-Zahlen größer als 4×10^6. Ingen.-Arch., Vol. 38, No. 4, 1969.

8. Nonweiler, T.R.F.: A New Series of Low-Drag Aerofoils. Aero. Res. Council., R&M No. 3618, 1971.

9. Liebeck, R.H., and Ormsbee, A.I.: Optimization of Airfoils for Maximum Lift. AIAA Journal of Aircraft, Vol. 7, No. 5, Sept - Oct 1970.

8

Afterword

by

Tuncer Cebeci

AMO has made significant everlasting contributions to the understanding and computation of aerodynamic flows. When the computing power was in its infancy, he recognized the importance of computers and pioneered in bringing them to the computation of aerodynamic flows in order to develop very beneficial methods for aircraft design.

His inviscid flow work for low speed flows, later extended to the calculation of complete flow fields around modem transport aircraft by Hess, is still one of the cornerstones of aircraft design methods. The boundary layer work, also initiated by him and later extended to separated flows by Cebeci, is one of the most effective approaches of incorporating viscous effects into inviscid flows. In recent years, this approach has been applied to AMO's subsonic inviscid flow method and has led to a unique powerful method for computing separated flows and predicting airfoil and wing characteristics. This approach is currently being applied to another powerful inviscid flow method for compressible flows developed by Forrester Johnson of Boeing. This method, known as TranAir, is the work horse of aerodynamic design methods used for many Boeing airplanes. Once the work is completed, it will provide a much needed capability for separated flows.

At this time, AMO's e^n-method is the only accurate, efficient, and practical method for predicting transition. Even though this method was proposed over fifty years ago, little progress has been achieved in this area towards developing better methods for predicting transition. Today, it is widely used for computing transition.

In the area of modeling turbulent flows, AMO developed a turbulence model which, with improvements by several researchers including Cebeci, became a very useful model for computing turbulent aerodynamic flows. At this time, this turbulence model is being extensively used in many calculation methods.

These are just a few examples of AMO's contributions to aerodynamics. His simple, basic insights that he infused into the calculation of aerodynamic flows and to the designs methods leave his indelible mark on future work in these areas.

Cheers!

9

Professional Biography of A. M. O. Smith

Publications

1. Flight Analysis of the Sounding Rocket. Journal of the Aeronautical Sciences, March 1938; F. J. Malina, co-author

2. The Jet Airplane Utilizing Boundary Layer Air for Propulsion, Journal of the Aeronautical Sciences, February 1947; H. E. Roberts, co-author

3. On the Motion of a Tumbling Body. Journal of the Aeronautical Sciences, February 1953

4. Improved Solutions of the Falkner and Skan Boundary-Layer Equations. Sherman Fairchild Fund Paper No. FF10, The Institute of the Aeronautical Sciences, March 1954

5. A Dust Method for Locating the Separation Point. Note in the Journal of the Aeronautical Sciences, April 1955, J. S. Murphy, co-author

6. Micromanometer for Measuring Boundary-Layer Profiles. Review of Scientific Instruments, August 1955, J. S. Murphy, co-author

7. On the Growth of Taylor-Görtler Vortices Along Highly-Concave Walls. Quarterly of Applied Mathematics, October 1955

8. Measurements of Wall Shearing Stress in the Boundary Layer by Means of an Evaporating Liquid Film. Journal of Applied Physics, September 1956, J. S. Murphy, co-author

9. Studies and Experiments in Drag Reduction by Boundary Layer Control: D.W. Clutter, co-author, Douglas Aircraft Company Rept ES 26354, July 1957

10. Transition, Pressure Gradient and Stability Theory. Proceedings IX, International Congress of Applied Mechanics, Brussels, Belgium, September 1956; Nathalie Gamberoni, co-author

11. Rapid Laminar Boundary-Layer Calculations by Piecewise Application of Similar Solutions. Journal of Aeronautical Sciences, October 1956

12. Exact Solution of the Neumann Problem. Proceedings III, U.S. National Congress of Applied Mechanics, Brown University, June 1958; Jesse Pierce, co-author

13. The Smallest Height of Roughness Capable of Affecting Boundary-Layer Transition. Journal of the Aero/Space Sciences, April 1959; D. W. Clutter, co-author

14. Analysis of Further Data on the Effect of Isolated Roughness on Boundary-Layer Transition in Supersonic Flow. Journal of the Aero/Space Sciences, January 1960; D. W. Clutter, co-author

15. Remarks on Transition in a Round Tube. Journal of Fluid Mechanics, Volume 7, Part 4, 1960

16. The Electrochemical Technique of Flow Visualization. Presentation Summaries, ASME Symposium on Flow Visualization, November 1960

17. Flow Visualization by Electrolysis of Water. Aerospace Engineering, January 1961; D. W. Clutter, co-author

18. Incompressible Flow About Bodies of Arbitrary Shape. IAS Paper No. 62-143, June 1962

19. Solution of the Incompressible Laminar Boundary-Layer Equations. AIAA Journal, September 1963; D. W. Clutter, co-author. In December 1983 I was asked by Science Citation Index to write a short article about this work because it has received very many citations. Published in Current Contents, Vol. 16, No. 2, January 14, 1985

20. Calculation of Potential Flow about Arbitrary Body Shapes. Proceedings International Symposium on Analogue and Digital Techniques Applied to Aeronautics; Liege, Belgium, September 1963; J. L. Hess, co-author

21. A General Method for Solving the Compressible Laminar Boundary-Layer Equations. Proceedings International Symposium on Analogue and Digital Techniques Applied to Aeronautics; Liege, Belgium, September 1963

22. Calculation of Nonlifting Potential Flow About Arbitrary Three-Dimensional Bodies. Journal of Ship Research, September 1964; J. L. Hess, co-author

23. Machine Calculations of Compressible Laminar Boundary Layers. AIAA Journal, April 1965; D. W. Clutter, co-author

24. A General Method for Calculating Low Speed Flow About Inlets. AGARD Specialists Meeting on Aerodynamics of Power Plant Installations, Tullahoma, Tenn., October 1965; J. L. Hess, co-author

25. General Method for Solving the Laminar Boundary-Layer Equations of a Nonequilibrium Dissociating Gas. AIAA Journal, April 1966; N. A. Jaffee, co-author; originally presented as AIAA Paper 65-129

26. Mass-Transfer Method of Measuring Wall Shear Stress in Supersonic Flow. Physics of Fluids, September 1965; A.R. Wazzan, R. C. Lind, co-authors.

27. Progress in Solving the Full Equations of Motion of a Compressible Turbulent Boundary Layer. Douglas Paper No. 3925, AD 634988, June 1966. Presented to the Seventh BOWACA Symposium on Aeroballistics; U.S. Naval Missile Center, Point Mugu, Calif., N. A. Jaffe, R. C. Lind, co-authors

28. Calculations of Potential Flow About Arbitrary Bodies. Progress in Aeronautical Sciences, Volume 8, Pergamon Press, 1966; J. L. Hess, co-author

29. Potential Flow About Two-Dimensional Hydrofoils. Journal of Fluid Mechanics, April 1967: J. P. Giesing, co-author.

30. Static Pressure Probes Derived from Supersonic Slender-Body Theory. Journal of Aircraft, September-October 1967; J. L. Hess, co-author

31. Solution to the Binary Diffusion Laminar Boundary-Layer Equations with Second-Order Transverse Curvature. AIAA Journal, September 1967; R. C. Lind, N. A. Jaffe, co-authors

32. Stability of Laminar Boundary Layers at Separation. Physics of Fluids, December 1967; A. R. Wazzan, T. T. Okamura, co-authors

33. The Stability of Water Flow Over Heated and Cooled Flat Plates. Journal of Heat Transfer, February 1968; A. R. Wazzan, T. T. Okamura, co-authors

34. Aerodynamics of Surface Roughness and Imperfections. SAE Paper No. 680198, presented at National Business Aircraft Meeting, Wichita, April 1968; Kalle Kaups, co-author

35. Solution of the Boundary-Layer Equations for Incompressible Turbulent Flow. 1968 Heat Transfer and Fluid Mechanics Institute, University of Washington, Seattle, June 1968; T. Cebeci, co-author

36. Recent Progress in the Calculation of Potential Flows. Proceedings of Seventh Symposium on Naval Hydrodynamics, Rome, August 1968

37. A Finite-Difference Solution of the Incompressible Turbulent Boundary-Layer Equation by an Eddy-Viscosity Concept. 1968 Symposium on Turbulent Boundary Layers, Stanford University, August 1968; T. Cebeci, co-author

38. Potential Flow and Boundary Layer Theory as Design Tools in Aerodynamics. CASI International Aerospace Exposition,. Montreal, November 26, 27; A. B. Bauer, J. L. Hess, co-authors, CASI Journal, February 1970 (Casey Baldwin Prize Paper)

39. Investigation of Heat Transfer and of Suction for Tripping Boundary Layers. Journal of Aircraft, September-October, 1968; T. Cebeci, co-author

40. On the Calculation of Profile Drag of Airfoils at Low Mach Numbers. Journal of Aircraft, November-December 1968; T. Cebeci, co-author

41. The Determination of Spatial Amplification Factors and Their Application to Predicting Transition. AIAA Paper No. 69-10, January 1969; N. A. Jaffee, T. T. Okamura, co-authors, AIAA Journal, February 1970

42. Calculation of Compressible Adiabatic Turbulent Flows. AIAA 2nd Fluid and Plasma Dynamics Conference, San Francisco, California, June 16-18, 1969; T. Cebeci, G. Mosinskis, co-authors. AIAA, November 1970

43. Recent Progress In the Calculation of Turbulent Boundary Layers. The Eighth U.S: Naval Symposium, Corona, California, May 6-8, 1969; T. Cebeci, G. Mosinskis, co-authors

44. A Decade of Boundary Layer Research. Applied Mechanics Reviews, January 1970

45. Solution of the Incompressible Turbulent Boundary layer Equations with Heat Transfer: Journal of Heat Transfer, February 1970; T. Cebeci, G. Mosinskis, co-authors

46. The Stability of Incompressible Flat Plate Laminar Boundary Layer in Water with Temperature Dependent Viscosity. 6th Southeastern Seminar on Thermal Sciences, April 1970, Raleigh, North Carolina; A: R. Wazzan, T. T. Okamura, co-authors

47. Laminar Boundary Layer Calculations on Bodies of Revolution in Hypersonic Flow. AGARDOgraph 147, 1970; G. L. Keltner, co-author

48. A Finite-Difference Method for Calculating Compressible Laminar and Turbulent Boundary Layers, Journal of Basic Engineering, September 1970; T: Cebeci, co-author

49. The Stability and Transition of Heated and Cooled Incompressible Laminar Boundary Layers. 4th International Heat Transfer Conference, Versailles, September 1970; A. R. Wazzan, T. T. Okamura, co-authors

50. Static Pressure Probes that are Theoretically Insensitive to Pitch, Yaw and Mach Number, Journal of Fluid Mechanics, Volume 44, 1970; A. B. Bauer, co-author

51. Calculation of Viscous Drag of Two-Dimensional and Axisymmetric Bodies in Incompressible Flows. AIAA Paper No. 72-1, 1972, or Journal of Aircraft, October 1972; T; Cebeci: G. J. Mosinskis, co-authors

52. Calculation of Laminar Boundary Layers by Means of a Differential-Difference Method-. Progress in Aerospace Sciences, Volume 12, Pergamon Press, 1972; N. A. Jaffee, co-author

53. Aerodynamics of High-Lift Airfoil Systems. AGARD Conference on Fluid Dynamics of Aircraft Stalling, Lisbon, 1972

54. Calculation of Separation Points in Incompressible Turbulent Flows; Journal of Aircraft, September 1972; T. Cebeci, G. J. Mosinskis, co-authors

55. Remarks on Methods for Predicting Viscous Drag. AGARD Conference on Aerodynamic Drag, Izmir, Turkey, April 1973; T. Cebeci, co-author

56. High-Lift Aerodynamics, 37th Wright Brothers Lecture, AIAA Paper No. 74-939, Journal of Aircraft, June 1975. The oral presentation was given 20 times around the U.S. and China.

57. Analysis of Turbulent Boundary Layers, December 1974, a book published by Academic Press; T. Cebeci, co-author

58. Remarks on Fluid Mechanics of the. Stall. AGARD-VKI Lecture Series 74 on Aircraft Stalling and Buffeting, March 1975

59. Stratford's Turbulent Separation Criterion for Axially-Symmetric Flows ZAMP, October 1977

60. Lecturer, jointly with Antony Jameson, Paul Rubbert, and Hideo Yoshihara for AIM Professional Study Series, Applied Computational Aerodynamics, Albuquerque, June, 1977

61. Tollmien-Schlichting Waves and Transition. Heated and Adiabatic Wedge Flows with Application to Bodies of Revolution. Progress in Aerospace Sciences, Vol. 18, No. 4., 1979; A. R. Wazzan, C. Gazley Jr., co-authors

62. H-R_X Method for Predicting Transition AIM Journal, June 1981; A. R. Wazzan, C. Gazley, Jr., co-authors

63. Optimum Tail Shapes for Bodies of Revolution, Journal of Hydronautics, Vol. 15, Nos. 1-4, January-December 1981; T. R. Stokes, R. S. Lee, co-authors.

64. The Boundary Layer and I, AIAA Journal, Vol. 19, No. 11, November 1981

65. Wing Design and Analysis - Your Job, Presented at 2nd Symposium on Numerical and Physical Aspects of Aerodynamic Flows. 17-20 January 1983, California State University, Long Beach. Also in Numerical and Physical Aspects of Aerodynamic Flows II, T. Cebeci ed. Springer-Verlag, New York 1984

66. The Possibilities of Laminar Flow Control, Soaring, December 1984

67. The Possibilities of Sailplanes Designed to use Laminar Flow Control, Technical Soaring, July 1985.

68. The Panel Method: Its Original Development. In "Applied Computational Aerodynamics" P.A. Henne, ed. In Progress, in Astronautics & Aeronautics AIAA, Vol. 125

69. Technical Evaluation Report on the Fluid Dynamics Panel Symposium on Aerodynamic and Related Hydroynamic Studies Using Water Facilities. AGARD-AR-225, August 1987

Addresses and Miscellaneous Oral Presentations

(Not resulting in externally published document)

70. Paper on Laminar Boundary-Layer Control. Symposium on Boundary-Layer Control, Washington, D.C., June 1953

71, Panel Member In Panel Discussion at Symposium on Boundary-Layer Control, Princeton University, Princeton, New Jersey, January 1954

72. Boundary-Layer Control, Principles and Promises. Aeronautics Seminar, California Institute of Technology, Pasadena, California, November 1955

73. Predicting Transition in Incompressible Flow. Institute of the Aeronautical Sciences Specialist Meeting, May 1955

74. Transition, Pressure Gradient and Stability Theory. Aeronautics Seminar, Massachusetts Institute of Technology, Boston, Mass., September 1956

75. Transition, Pressure Gradient and Stability Theory Edwin Baetjer Colloquium, Princeton University, Princeton, New Jersey, September 1956

76. Effect of Isolated Roughness on Transition in the Laminar Boundary Layer. Aeronautics Seminar at Pennsylvania State University, Pennsylvania, March 1958

77. Exact Solution of the Neumann Problem. Lecture to David Taylor Model Basin, Carderock, Maryland, June 1958

78. Effect of Isolated Roughness on Transition in the Laminar Boundary Layer, Institute of Aeronautical Sciences Specialist Meeting, Los Angeles, California, October 1958

79. Flow Visualization Using Water as the Working Medium. Institute of Aeronautical Sciences Specialist Meeting, Los Angeles, California, September 1959

80. Hydrodynamic Research at Douglas Aircraft Company. BOWHAC Meeting, Pasadena, California, March 1961

81. Systematic Design of Improved Static Pressure Sensing Probes. BOWACA Symposium, Silver Springs, Maryland, October 1961; J. L. Hess, T. L. Rivell, co-authors

82. On Machine Calculation of Boundary Layers. Fluid Flow Symposium, Georgia Tech., April 1967

83. Numerical Solution of Laminar Boundary Layers. AGARD Seminar on Numerical Methods for Viscous Flows, National Physical Laboratory, Teddington, England, September 1967.

84. Progress in Calculating Turbulent Boundary Layers and Separated Flows, Purdue University, March 1968

85. Fluid Mechanics - Bridging the Gap Between Science and Engineering Design. Seminar at University of California, Berkeley, California, April 1969

86. Short Course in Boundary Layer Stability, UCLA, June 1969

87. Seminar on Calculation of Turbulent Boundary Layers, NSRDC, July 1969
Seminar - Fluid Mechanics - Closing the Gap Between Science and Engineering Design. Caltech, November 1969

88. The Douglas Neumann Program. NSRDC Accomplishments Seminar, February 1970

89. Seminar on Calculating Boundary Layers. University of Cincinnati, January 1971

90. Seminar - Fluid Mechanics-Bridging the Gap Between Science and Engineering Design. Naval Ordnance-Laboratory, April 1971

91. Seminar on High-Lift Aerodynamics; Purdue University, November 1972

92. Briefing on Transition to Jason Study Group of DARPA, April 1974

93. The e^{n} Method with Special Reference to Bodies of Revolution. Rand Corporation Transition Workshop, July 1974

94. The Aerodynamicist's Bag of Tricks. Presented at NASA Colloquium Series, December 1974. Also to NSRDC, January 1975, Northrop Inst., May 23, :1978

95. Seminar on Low Speed Boundary-Layer Transition, UCLA, March 1976

96. Rand Transition Workshop, September 15, 1976, report on work at UCLA - Navy Contract

97. Seminar on Laminar Flow control, Stanford University, January 1977, and C4T, May 1980

98. First Stillwell Memorial Lecture - Laminar Flow Control, Promises and Problems, April 1980, Univ. of Illinois

99. The Possibilities of Sailplanes Designed to Use Laminar Flow Control. Presented bo Southern California Soaring Association, June 20, 1984. Also presented at National Meeting of Soaring Society, San Diego, March 1, 1985.

100. UCLA Current Challenges in Aviation, January 25, 1986.

101. Reminiscences on the U.S. Naval Technical Mission in Europe. AIAA Ventura-Pacific Section, Oxnard, April 21, 1993.

102. Reminiscences on the U.S Navy Technical Mission to Germany at the End of World War II. AIAA San Gabriel Valley Section, March 16, 1995

Honors

Robert H. Goddard Memorial Award for Liquid Propellant Rocket Development, American Rocket Society, 1954
Engineering Achievement Award, Douglas Aircraft Company, Inc., El Segundo Division, 1958
Elected Fellow AIAA, 1969
Recipient (Joint) of Casey Baldwin Award of CASI, 1971
Wright Brothers Lecturer - AIAA, 1974
D. Sc. Degree, (Honorary) University of Colorado, May 1975
Fluids Engineering Award, Amer., Soc. of Mech. Engineers, 1985
Elected Honorary Fellow, AIAA, 1986
Elected to National Academy of Engineering, February 1989

Miscellaneous Memberships and Activities

- Member, United States Naval Technical Mission in Europe, Paris, May-Aug., 1945
- Visited most major European Aeronautical Research Establishments in connection with problems of boundary-layer control; U.S. Navy sponsorship, January-March 1955
- Boundary Layer Theory, University of California at Los Angeles, 1954-1958

- Technical Reviewer for the Journal "Applied Mechanics Reviews," 1957-1983
- USAF DAG Committee (Division Advisory Group) a special committee formed at times to advise the Air Force on a difficult problem, 1964-1965
- Session Chairman, ICAS, Paris, 1964
- Session Organizer, ASME Annual Aviation and Space Division Conference, Los Angeles, June 1968
- Session Chairman, AIAA Fluid and Plasma Dynamics Conference, Los Angeles, June 1968
- Continuous activity as referee for AIAA, ASME, National Science Foundation, various symposia and congresses concerning papers and paper selection
- Member, discussion panel, Symposium on Compressible Turbulent Boundary Layers, Langley Research Center, December 1968
- Session Chairman CASI/AIAA Subsonic Aero- and Hydro-Dynamics Meeting, Ottawa, Canada, July 1969
- Member, Ad Hoc Committee of USAF Scientific Advisory Board, 1971
- Chairman of Discussion Panel, Project SQUID Symposium on Fluid Dynamics of Unsteady Three-Dimensional and Separated Flows, Georgia Tech, June 1971
- Session Chairman, G.M. Symposium on Drag of Automobiles, September 1976
- Expert witness in a patent infringement suit involving aerodynamics, May 1977
- Session Chairman, USC Symposium on Applications of Computer Methods in Engineering, August 1977
- Member, AIAA Fellows Selection Committee, March 1978
- Discussion Panel Member - Coherent Structure of Turbulent Boundary Layers, Lehigh University, November 1978
- At the invitation of the Peking Institute of Aeronautics and Astronautics, he was in Peking, China, from September 7 to November 12,1979, giving a series of lectures
- Chairman of Discussion Panel, Numerical and Physical Aspects of Aerodynamic Flows, California State University, Long Beach, CA, January 1985

Memberships

Sigma Xi, California Institute of Technology
AIAA (Fellow)
NACA Subcommittee on Internal Flow, 1948-1951
Editorial Advisory Board, AIAA Student Journal, 1963-1972
American Physical Society
American Association for the Advancement of Science
American Defense Preparedness Association
American Society of Mechanical Engineers
Registered Mechanical Engineer - State of California, License No. 2554

Who's Who in Engineering
Who's Who in Technology Today
Who's Who in the West
Men of Achievement
American Men of Science
Personalities of the West and Midwest
Engineers of Distinction
Who's Who in California
Who's Who in Aviation and Aerospace
Leading Consultants in Technology
International Who's Who in Engineering
Who's Who in Frontier Science and Technology
Heat Transfer Panel, Bureau of Naval Weapons Advisory Committee for Aeroballistics
(BOWACA) 1964-1967
AIAA Fluid Dynamics Committee, 1966-1969
Heat Transfer Panel, Naval Aeroballistics Advisory Committee (NAAC), 1967-1971
Board of Editors, "Progress in Aerospace Sciences," Pergamon Press, 1968-1975
ASME K-12 Committee on Aircraft and Astronautical Heat Transfer, 1969-1972
American Editor for "Computer Methods in Applied Mechanics and Engineering," 1972 to 1977
Member of Selection Committee for Freeman Scholar Award of ASME, 1976-1982
Member of DOT CARP Committee on Auto Aerodynamics, January 1980
Member NASA Aeronautics Advisory Committee (AAC), 1980-1983
Member HLDV Committee of Naval Underwater Systems Center (NUSC), 1981-1982
Who's Who in America, 1990
Marquis Who's Who in the World, 10th Edition
5000 Personalities of the World, American Publishing Who's Who
Marquis Who's Who in Science & Engineering, 1992, 1st Ed.
Strathmore's Who's Who Registry of Business Leaders, 1995
American Men and Women of Science, 1995,1997
Directory of Distinguished Americans Directory of International Biography